"*ReCo2gnition* is an intelligent, well-crafted thriller with great forward momentum, lots of intrigue and memorable characters."

"What happens when Science Fiction meets science, art, philosophy, and architecture? Well, naturally, you get a compelling story! Dowson has outdone himself in this second book of the Recognition series. The action ramps up while still delivering insight all designed to help us of this time to avoid mistakes which will make the future wholly unlivable. Thought provoking and inspiring!"

"Science fact and science fiction collide in this eco-thriller from Mark Dowson. Two time travelers from 2112 return to the present day, one to assassinate a wind energy engineer, the other to rescue him—and perhaps save the planet!"

"A convincing portrayal of a potentially dark future, a clear warning that our current lack of commitment and reckless environmental behaviour can destroy nature and end our dream of a prosperous future.

"The vision and conviction of the author is not only outstanding but serious. The shorty line is effective in exploring new thinking and even technical solutions.

"It is promising to see that the new generation is taking environmental challenges seriously and trying to spread the idea through story telling."

"If you like crime, mystery and romance all wrapped up in a sci-fi thriller, this is for you. Mark has also introduced today's global concerns as a crucial theme, and if that isn't enough, he's cleverly focused on mental health too. You'll need to concentrate, but that's its magic. This, the middle of Mark's trilogy sets things up nicely for the third book finalé.

Once again it's a fast-paced read full of thought provoking science based drama with just a touch of sci-fi to embellish the storyline.

It's up to date making use of the Coronavirus and yet still manages to draw the reader into the future and the past.

You need to concentrate - but it's worth it."

— GRAHAM MILLER, Public Relations Executive at Media-Vu, Media Consultant, Broadcaster at BBC News, Public Relations Advisor, Media trainer

"A great sequel from Mark Dowson to his first instalment of *ReCO2gnition*. Building on the enigmatic characters of the launch volume of *ReCO2gnition* the author maintains his pursuit of exploring big contemporary challenges to the human condition against the intriguing backdrop of futuristic science fiction. This time it is the pandemic. Another intriguing and enjoyable read."

— DR. VOLKER BUTTGEREIT, CDir, PhD, DIC, M.Eng, ACGI, Managing Director, FD Global Limited, 'Designing Large Scale Building-integrated Wind Turbines for the World Trade centre in Bahrain'

"Sci-fi, ecosensitive, time-travelling thriller which alternates between the present and a post apocalyptic world where a leading scientific genius must be transported to the future to save humanity. Keeps you guessing to the end whether he will succeed or not."

— JAMES TUTTLE, PhD in exercise Physiology. Teacher of science. Elite distance runner

"A great read. Time shifting and hi technology giving a vision of the 22nd Century if we let things continue as they are, not a pleasant vision of the future. Our thoughtful lead, Ben Richards leaves all the super hero stuff and the 22nd Century players. Very like a Greek mythical legend would go on a journey of discovery to save the future of the planet. The environmental issues of global warming are well discussed and the consequence of bad decisions clearly a warning for all readers. Each of the characters are well fleshed out and all the tech detail felt credible."

—ALASDAIR GIBSON, Sales Director, Mohn Media UK

"Creative Mark Dowson is at it again, drawing your mind into moments of imagery and situations that bring characters to life."

— ADRIENNE MAZZONE, President, TransMedia Group

"I read book 1 and couldn't wait for book 2. You won't be disappointed. Book 2 is even faster and more thrilling! You don't want to put the book down. Mark Dowson with his expertise on wind energy has created a Sci-Fi Thriller, no one else can! An incredible read that truly makes you think about our world."

—ANNA MARIE PELOSO, Internationally acclaimed Narrator of audiobooks, New Jersey, USA.

"Groundbreaking time travel concept that is original, revolutionary and has never been thought of before, in the world of science fiction. Mark Dowson uses a factual artefact as mystical form of timetravel. Fascinating!"

—KARTIK SPRATT, Director, Thames Consultancy Services.

"Doctor Richards is back and time travlling and mystery solving has never been so engaging. Ben Richards is pulled further into a future conspiracy...but has time ran out for him and us? A thoroughly entertaining read this ground breaking sequel picks up shortly after the original and looks even deeper into the millenia old mysteries of sator square,"

RECO₂GNITION:
Co-Anda 19
Part 2

by **Mark Dowson**

IISBN 978-1-9168957-0-6

Published by

Abstract2Construct Limited

38 Wenning Lane,
Emerson Valley,
Milton Keynes,
England
MK42JF
www.markdowsonauthor.com

Mark Dowson

RECOG$_2$NITION

CO-ANDA 19 VACCINE
PART 2

MARK DOWSON

PROLOGUE

The blades rotated rhythmically.
Patiently.
Methodically.
I was at peace. Comfort and reassurance was all around.
Darkness crept over me. Then there was a pale light. The
blades were still there. Always four. One. Two. Three. Four.
An unchanging shape in front of me. Or above me.
The same, regular rotation.

But, what are the blades?
Where are they?
Why are they there?
Suddenly, I feel my entire torso jolt.
He hears the words..."Let her go, she is gone, she is gone."

CHAPTER 1

SHUI FENG

Shui Feng made his way to the driver's side of the white Alfa Romeo, speaking into a communication device on his wrist, as he went: "We now have them both. Take Richards' female companion to Pompeii, as arranged. You have done well, TX1. I will analyse your data more fully later, but it appears we will focus our attention on the ruins of Pompeii. I would not be surprised if there is a portal in that location."

With that, he fired up the engine of the Duetto. It roared into life, but Feng calmly and slowly pulled away and eased the car into the traffic that circulated around the piazza.

CHAPTER 2

MERISI

Merisi stood in the doorway of Caravaggio's restaurant, surveying the scene. All was quiet, and the rain that had momentarily been torrential had now eased to become a steady, monotonous shower. There was no sign of the struggle that had taken place in the street, just seconds earlier – except for the faint glow of a smartphone's display, which Merisi could just about make out, in the gutter, very close to a drain, on the opposite side of the street.

Merisi pulled his white panama hat further over his eyes as he ventured into the rain. With the collar of his white jacket turned up, so that just a few locks of wavy, black hair were revealed, he peered out, looking from side to side, from beneath the brim of his hat, as he crossed the street. He picked up the 'phone, whose display was all but extinguished, and sighed. His worst fears had been realised.

He put his hand into his inside jacket pocket, and, without fully revealing his communicator, said: "It's Merisi – Richards has gone. The only trace of him is his phone, which he was using to speak with me. I don't know exactly what has happened, but I fear the worst... Yes, I administered the drug, but I'm not sure how much it can help him now .... No, I don't think that my cover has been compromised, but that's less important now. I'm afraid I lingered in the shadows for too long. I put too much faith in the secrecy of our mission at stake here... Yes, I know... I don't know how much GIATCOM knows – but, if they have Richards, and given that Feng is involved, I think we must assume that they will find out pretty quickly... Pompeii? Again, I think we need to assume that they will know about Pompeii soon enough, so they will be there, before long... Yes, I agree... Yes... I am on my way – I will report to you as soon as I arrive."

Merisi removed his hat and ran his hand through his mop of black hair. He drew a deep breath, surveyed the scene around him one last time, and set off in the piazza's direction.

CHAPTER 3

FEAR OF THE UNKNOWN

AUGUST 12, 2017

"*N*othing *in life is to be feared. It is only to be understood.*" – Madame Curie*

The early morning sea swoops low along the horizon, revealing a sweeping panoramic view across a vast and calm tropical ocean. Gentle waves lap over a woman's shoulders as they glisten in the sun. She floats confidently in this body of water, albeit a fair distance from the shore. In a peripheral view, an animated young boy is leaping up and down along the shoreline, trying to catch her attention. The focus is then drawn to the child as the woman remains oblivious to him. He is around 12 years old and wears a silver-grey T-shirt with a dark blue logo which reads 'UNA.'

The woman continues to float, unawares of the young boy. She is now on her back with her long dark locks of hair splaying around her shoulders just beneath the waterline. She looks intensely relaxed.

The young boy flails his arms frantically, shrieking as he looks out to sea.

"Muuuuuuuum!"

His cries continue to go unheard by her. The beach is deserted. Numerous towels, piles of clothes and strewn belongings remain. People had clearly left in a hurry.

Day light fades to darkness.

* * * * *

The view switches.

I am on my back, moving quickly and looking upward. My senses and vision are obscured, I cannot see or hear anything for a lingering moment or two. A dull, intermittent sound draws close, gathering volume, louder and louder as we slowly come to my senses. The contrast between darkness and light fades in and out, as my vision blurs and flashes on and off. I can just about make out swishing blades that revolve and flicker above my head in my distorted field of tunnel vision. I can't quite tell if the muffled sound and vision is that of a helicopter rotator blade. My vision fades out into total blackness.

* * * * *

The view switches.

A view of what resembles the planet Earth from space appears. But as we zoom in to view the Earth more closely, a closer view reveals a montage of risen sea levels, devastated cities, decimated countries and drastically altered continents, entire populations wiped off the face of the Earth...a major global disaster has clearly occurred. The view is then pulled back, ascending into the brown and dusty sky which now surrounds the planet. The montage of devastation ends. Once again, I hear the muffled sounds of the rotor blades. Suddenly my vision is blurred by a brilliant white light which completely consumes my field of vision.

* * * * *

Dr Ben Richards opened his eyes. Transfixed by the rotor blades of the ceiling fan rotating above, he felt dizzy and sick. The humming sound of the rotors perpetuated like a ticking clock. The ceiling fan tried its best to send him into a trance. His head was throbbing. He closed his eyes and then opened them again. He contemplated the white ceiling. It was made up of white plastic panels, each speckled with random black dots. He lay there for several seconds, examining the randomness of the dots, and drawing imaginary lines between them in his mind.

He closed his eyes and then opened them again. He contemplated the white ceiling. It was made up of white plastic panels, each speckled with random black dots. He lay there for several seconds, examining the randomness of the dots, and drawing imaginary lines between them in his mind.

He felt desperately nauseous, and he could not escape the pain in his head. He blinked twice – then twice more. Very soon, a new feeling came over him. He now felt disorientated, and, as consciousness developed, wondered where he was.

What time was it? What day was it? And, more importantly, what was he doing in that bed?

He lay there for a moment more, while he tried to recall how he got there and tried to identify his symptoms and recall how he caught these symptoms. "Was it a virus?" "Who did he catch it from?" he wonders, and then suddenly remembers after seeing a glimpse of the Italian lady talking to the doctors in the corridor through the half-opened door in the hospital room, that it was her. He recalled being alright until... he remembered Merisi on the phone saying he had slipped something into his drink at Carravagio's restaurant. A likely increased dose of the Co-anda19 vaccine, which would have been the likely cause of enhancing his senses to recognise there was something not quite right with the Italian Lady. But it couldn't be that drug that made him weak, nausea, headaches, and difficult to breathe, he felt incapacitated. It must have been the spray he remembers the Italian Lady spraying in his face. Could Merisi have known... perhaps? That he was about to be captured, so in advance, gave him a drugged drink so that he could identify the Italian lady before his capture? Whatever the reason was to drug him, to make him feel this way, the Italian Lady was not who she seemed to be. And neither were her intentions.

The chemical he inhaled was clearly MANMADE, and in the wrong hands was lethal, he thought.

Richards raised his head with a jerk– and immediately, the pain in his head intensified. After squinting momentarily, his eyes swiveled in their socket, as he anxiously surveyed his surroundings. He peered through the door into the corridor again and could see and hear two doctors in white coats standing and talking right outside.

"If there is an overrun of airborne viruses in this hospital, we will have a global pandemic", one of the doctors explained to another doctor. "Yes, we will have Oxygen Debt! Oxygen displaced with an increase in airborne viruses."

"Which will mean we will need to circulate more air changes, if more people in a space is required, which will cause more oxygen displacement. This new natural ventilation system installed purely for energy efficient reasons to avoid higher CO_2 emissions than mechanical ventilation is not a good idea. It is compromising health and safety. Surely it's better to have mechanical ventilation so quicker air changes occur to avoid oxygen depletion in the air from an increase in airborne virus."

"Totally agree Doctor, we need more air changes within any given space with the hospital and natural ventilation will not give us that. So, we avoid putting everyone at risk of being exposed to Oxygen debt in any room within the hospital, caused by these viruses."

"Increased air changes are the key to avoid oxygen debt, so we need to rethink the sustainability and energy efficiency design in buildings. It's all wrong!"

Richards sighed as he thought the topic of discussion was most interesting, and something that he could use for inspiring a new design philosophy.

He was alone, in a small room with magnolia walls. These walls were bare, and there were no windows. There was a chair on either side of the bed, and a Spartan-looking, charcoal grey two-seater sofa in the corner of the room, diagonally opposite the point from where he viewed his predicament.

As he engaged his arms and his legs, and tried to raise himself up, he became aware of a plastic peg that was clamped onto the index finger of his left hand, and then he noticed a half-full glass of water on a table, to his left and just above the level of the bed's mattress. Beyond that, he saw a machine with a round, green screen in its centre, with a bright, squiggly line moving silently at regular intervals across it. There were many wires leading from this machine.

These were all the unmistakable trappings of a medical institution. He was in a hospital. But, where? And, more importantly, why?

With an effort, he raised himself to a seated position, and placed his pounding head in his hands. At that moment, he heard the rhythmical click-clack of stiletto heels approaching the room – and soon he was hugely reassured to be greeted by a familiar, smiling face!

"Grazia!" he cried.

"Benjamin! You are awake!" Grazia hurried towards him and grasped both of his hands as she sat on the bed beside him. Richards immediately felt more relaxed. In the midst of all the confusion, the one thing that he felt he could be sure of was Grazia's friendly face, and her lovely smile.

"Grazia! What's going on? What's happened to me? Why am I in hospital?"

"Don't worry. You have had an accident, and you bumped your head – but you are going to be fine."

"But – but I don't remember having an accident. And where the hell am I?"

"It's OK, Benjamin. The head trauma has caused a slight memory loss – but it will soon return."

Suddenly, it struck Richards that it was very strange that Grazia had addressed him as 'Benjamin'.

"'Benjamin,'" he exclaimed, with an uncertain smile.

"What?"

"'Benjamin.' You called me 'Benjamin.' I'm 'Ben.' My mother is the only person who has ever called me 'Benjamin.'"

Shui Feng silently castigated himself for his error.

"Yes. 'Ben.' Yes, of course. I was only joking. But never mind that. The doctor has given me a very important job to do. He has instructed me to talk to you, to help you to regain your memory."

Feng made an effort to smile his most reassuring smile. "Just relax – Ben – and try to remember," he added, as he gently pushed Richards' shoulders back, so that, once again, he was lying with his head on the pillow, staring up at the white ceiling tiles.

Richards wracked his brain for any recollection at all, but he was aware of nothing more than the thumping pain in his head.

His desperate introspection was interrupted by the sound of two people having an animated conversation in the corridor. They were speaking words he could not understand, but the musicality of the language was unmistakably Italian.

"Italian," he muttered. "Italy. Yes, I had to go to Italy."

"Yes. Go on..."

"I had to go to Italy – to Milan." Then, he was hit by a realisation. "I am here for a seminar. Yes – I am going to present at a seminar!"

Richards experienced a rush of anxiety. Suddenly, his mind was engulfed with memories of a thousand dreams that he had had, each one of them based around the theme of being late for a presentation that he was due to give, or of losing his speaking notes, or of becoming lost on the way to the venue...

"The seminar! The seminar! I am here, in Italy, to give a presentation at the seminar." He immediately sat up and grasped the bed sheet, ready to cast it to one side and dash out of the room. He was halted by a warm, sun-tanned hand on his chest, pushing him softly, but firmly, back down again.

"No, Ben. Everything is fine. You must just relax."

"But – my presentation!"

Richards made another desperate attempt to climb out of the bed – and, this time, he was prevented from doing so more forcefully, as Feng put a hand on each shoulder and thrust him back down, with a display of physical strength which took Richards by surprise.

"No, Ben," said Feng, with a fixed stare that was insistent to the point of being menacing. Then, as if a switch had been flicked on, Grazia's kindly expression returned, and her eyes were smiling fondly at Ben once again. "No, Ben. Everything is fine. You delivered your presentation the day before yesterday. Do you not remember? Do you remember nothing of the seminar?"

Gradually, Richards started to regain some vague memories of standing up in front of an audience and delivering the presentation that he'd reeled off on a number of occasions.

"Well," he began, with a confused frown, "sort of. Yes, I remember the auditorium. A large room, with a high vaulted ceiling."

"Who did you meet there? Do you remember meeting anybody?"

Richards struggled to recall any details at all from the seminar. "No, Grazia. Not really."

"Then, how about afterwards? We went for a nice meal together, didn't we? Don't you remember? We went to that nice little restaurant – called Caravaggio's – didn't we?"

"A little restaurant? An Italian restaurant?"

"Yes."

"What did I have?"

Feng hesitated. His knowledge of Italian food from the period was limited, to say the least.

"Pasta. Yes, you had pasta."

"And what about you? What did you have?"

"I also had pasta," exclaimed Feng, who was becoming increasingly impatient. "But, do you remember it? The restaurant!"

"Yes. Yes, I think I do."

Suitably encouraged, Shui Feng became more focused with his questioning. "And do you remember meeting anyone there?"

Richards stared back at Grazia, with a look of confused blackness.

"Ben – do you remember you met a friend of yours, in the restaurant? What was his name?" Feng asked himself, pretending to be searching his memory for a person's name. "Ah! Merisi! That's it – Merisi. You met your friend, Merisi, there, didn't you? You remember? The man with the beard."

"Merisi. Merisi," repeated Richards. "No, I don't know anyone called Merisi," it was also a very truthful response that he had met nobody with a beard, given that Merisi had maintained his disguise as the old man whilst in Caravaggio's.

Feng rose to his feet and paced impatiently to the door and back again. When he returned to Richards' bedside, he saw that the young engineer's eyes were closed, and that he had returned to a deep sleep.

* * * * *

As soon as Ben Richards was awake, Shui Feng sprang from his seat where he had been sitting, watching and listening intently as the man slept. He leant over Richards, whose face and neck were now glistening with perspiration, and looked into his eyes.

"Ben, you are back!"

"Back?"

"Yes, you were sleeping – and, it seems, dreaming."

"Yes, I was," muttered Ben, flicking glances anxiously about him. He looked up at the ceiling and seeing the white tiles confirmed to him he was still in the hospital.

"What is it you saw, Ben – in your dream?"

"I saw many things. It was a dream the like of which I have never had before. Everything seemed... well... so real!"

Feng took Richards' hand once again. "Tell me, Ben. Tell me what was in your dream."

"It is not easy to describe," began Ben. "So much seemed to happen, and all at once, it seemed."

"Try. Try to describe it. It will help with your amnesia," implored Feng, with the most reassuring smile he could manage.

Richards tried to recount everything that he could remember from his dream, but it was when he told Grazia about the mysterious strangers who appeared to be warning him of grave and imminent dangers, that Feng's interest perked up.

"And who do you think these people were, Ben?" asked Feng, who was attempting, unsuccessfully, to feign just a passing interest in their identity. "Do you think that one of these men might have been your friend, Merisi?"

"Merisi? No, I don't know anyone of that name. Why do you keep asking me about him?"

"Oh, it's just that you kept saying his name, while you were asleep," lied Feng. "Describe these men to me."

"Well, one of them was an old man. Very short – certainly no taller than you, Grazia. The other one was a few years older than me; I suppose. He had a small, black beard, and long black wavy hair."

Feng immediately recognised his old foe. He leant closer to Richards. "And what, precisely, did this man tell you, in your dream?"

"A load of fanciful nonsense, it seemed to me. A lot of stuff about volcanoes and time travel and some sort of nuclear catastrophe."

"Anything else?"

"Well, both of these men, in my dream, kept emphasising the importance of certain words to me."

"Words?" asked Feng.

"Yes. Foreign words, that I didn't understand – or maybe they were part of a code, or something. I remember that 'SATOR' was one of them."

Feng was now convinced that he was onto something.

"It was all very weird – yet so real," continued Richards.

"Does 'SATOR' mean anything to you? Is it an Italian word?"

"No, it's not Italian."

Richards sat up, with an effort, in the bed.

"I'll tell you what I really need, Grazia," he said, still squinting in reaction to the headache, which had worsened. "I could do with a visit to the loo, and a cup of tea – in that order!

My mouth feels like the bottom of a parrot's cage!"

"Yes, of course," said Feng, with a smile. "There is a toilet just across the corridor, and I will go and get you something to drink."

Richards turned his left cheek towards Grazia, in the hope that he might get a kiss – through sympathy, if nothing else – but, in an instant, Feng was gone, and with some urgency, leaving the sound of hurrying stilettoes echoing in the corridor.

CHAPTER 4

THE CALL
TO MERISI

Wearily, Richards swung his legs out of the bed, placing his feet on the cold, pale grey ceramic floor tiles. He was shocked to find that he was dressed in nothing but a backless surgical gown, which did little to preserve his dignity. He saw his clothes were piled in a heap on one chair in the room, so he removed the plastic monitoring peg from his finger, and made his way, stiffly at first, to the chair. He stooped to pick up his shirt and his trousers, and, as he did so, two small coins, and what looked like a business card, felt onto the hard floor, with a 'clank'.

Richards picked up the card, and it immediately seemed familiar to him. On one side was a SATOR SQUARE; on the other, there was a miniature print of a Baroque painting, and, within a thin margin at the side, the handwritten name "MERISI," followed by a telephone number.

"Merisi!" he said to himself. So, there was some truth in the names and images that had been swimming around in his head. It appeared, also, that there really was a person called Merisi, and it was easy to believe that he was a friend of his, as 'Grazia' had suggested.

Richards became very excited about calling the number on the card and couldn't wait to tell Grazia about his discovery when she returned. He searched his trouser pockets, and then. his other clothing for his mobile 'phone – but it was nowhere to be found He then noticed an avocado-coloured, wall-mounted 'phone near to the doorway of the room. Instructions for making external calls were in Italian, but he guessed he would need to dial "0" – and then punched in the number that Merisi had left him. The call was answered almost immediately.

"Pronto?"

"Err – yes. Is that Merisi?"

There was a brief pause – and then: "Dottorre Richards! Is that you?"

"Yes, it is!"

"Thank goodness! I thought I'd lost you!" cried Merisi. "And Signorina Rossini? Is she safe, too?"

"Yes, she's safe and well! Grazia's here, looking after me. You're not going to believe this, but, apparently, I have had some sort of accident. I banged my head, and it has left me with a degree of memory loss – so I honestly don't know what happened to me."

"And the young lady is absolutely fine?"

"Yes, she is unscathed. We are both fine. She is helping me to try to get my memory back. In fact, it was Grazia who told me you and I are friends. Otherwise, the name 'Merisi,' which I found on a card you gave me, would have meant nothing at all to me. This amnesia is really annoying. The sooner I can remember what I'm supposed to be doing here, the better."

"'Amnesia,' you say?" asked Merisi, now sounding more guarded.

"Yes, I remember saying that – are you trying to test me?" said Richards, now feeling far more cheerful, now that he had connected with someone familiar, even if he had not known who Merisi was, just seconds ago – but Merisi remained troubled at the suggestion of memory loss.

"So, what exactly can you remember, *Dottore*? Do you remember making a presentation at the Ministerio di Energia? And do you remember going to Caravaggio's restaurant, in the evening?"

"Well, I can just about remember going to the seminar, and getting up to make a speech – but I recall nothing at all after that."

There was a pause.

"You began your presentation approximately fifty hours ago. Tell me, *Dottore* Richards. Have you been sleeping rather a lot... and very deeply? And have you been waking up with your mouth feeling very dry, especially the roof of your mouth?"

"Well, yes."

"And I don't suppose that there is any external sign of any injury to your head? No lump? No stitches?"

Ben Richards ran his free hand across his scalp, in all directions.

"Actually, now that you mention it – no, there's nothing."

"And what have the doctors there told you about your condition?"

"Well, to be honest with you, I haven't yet seen a doctor. You see, I've been sleeping for most of the time that I've been here, and Grazia says-"

"Ah, Grazia! No doubt she has been asking you many questions about me – and no doubt she wants to know all about what you learnt in the restaurant?" Merisi took Richards' shocked silence as confirmation of his worst fears. "*Dottore*," he said, with a sigh, "You must get out of there!"

"But-"

"This will be difficult for you to understand, but you have been drugged. The last I knew of you and your lady friend is that you had both been taken by some very dangerous people whose capabilities are well beyond your comprehension. These people are very good – and there is no way that you would be able to escape them if they didn't want you to escape."

"Oh, no! Grazia! She has just popped out to get me a cup of tea! I must find her and warn her-"

"No! No! Grazia is not who, or what, she appears to be. This is something else that you must trust me on, *Dottore*. You must LEAVE NOW! – And don't let 'Grazia' see you."

Richards was unable to respond immediately. All of his

attention was focused on the dilemma that was now right in frontof him. Should he believe this man, who, a short time ago, he thought he had never heard of? Could he leave Grazia and abandon her to an uncertain fate?

He clasped his hand to his forehead.

"*Dottore*! Are you still there?"

What dictated Richards' next course of action was partly the genuine concern for him that there appeared to be in the man's voice – but it was mainly a strong sense of déjà vu that he felt on hearing the words 'LEAVE NOW.'

"This is not the first time that you have been there to save me, is it?"

"No, *Dottore*. It isn't – but, if you don't get out of there now, you will be beyond my help forever!"

That was enough to convince Richards.

"Okay. So, what do I do? I don't even know where I am", he said, anxiously looking all around him. "Hang on. It says above the 'phone: '*Centro Cardiologico Fondazione Monzino*,'" he read, slowly, and with some difficulty.

"I will find it. Just get yourself to the main entrance and wait for me there. Do you even know what floor you are on?"

"No, but I know there is a lift at the end of the corridor. I have heard it."

"Good! Get into it now, and don't, under any circumstances, stop for anyone – and keep well away from Grazia Rossini!"

CHAPTER 5

THE MEETING
WITH MERISI

Ben Richards peered out through the leaves of an ornamental bay laurel that had been planted adjacent to the entrance of the hospital. It was dark, but Richards had a good view of the large car park and the approach to the building. All seemed very quiet, with just the occasional vehicle passing by on the main road in the middle distance. The early hours of the morning were approaching, as began to feel more than a little chilled. After putting his shirt and trousers on in a hurry before leaving the hospital, he realised he had left his body exposed to the falling air temperature. He shivered – but guessed that this was more to do with shock and with the trauma that he had recently undergone, than due to the temperature of the night.

As he waited, a sense of anxiety grew. What had at first begun as a curious sense of disorientation was now being replaced by a very real feeling of uncertainty and vulnerability. If, as seemed likely, he had been drugged, then who could have done this... And why? Who could he trust? Why was 'Grazia' now such an untrustworthy character? Who was she?

The only thing that Richards felt certain about was that this person called 'Merisi' was somehow on his side – and he would have plenty of questions for him when he eventually appeared. But what did Merisi look like? Could he rely on Merisi recognising him?

Richards broke off from his hidden vigil when his aching head could tolerate the glare from the hospital's neon signs and the nearby streetlights no longer. He averted his gaze from the light-bathed scene and examined the card that he had hurriedly put into his trouser pocket as he left the hospital ward.

He thought hard about the SATOR SQUARE, and the name 'Merisi,' but it was no use. The only memories they stirred were of the dream from which he had recently awoken – and those memories were fading fast.

Richards turned the card over and scrutinised the miniature reproduction of the painting that was printed there. He recognised the painting as being one by local artist Caravaggio, but it otherwise meant nothing to him.

His inspection of the card was interrupted by a bright blue light, which came from the other side of the laurel bush. Instinctively he crouched low, held his breath, and watched and waited. Soon, he heard footsteps approaching. It was the measured step of a man; he thought – certainly not the urgent click-clack of Grazia's stilettoes.

Richards began to rise from his crouch, in an effort to catch a glimpse of who it might be. As he did so the footsteps stopped, and he could detect through the dense foliage the figure of a man dressed from head to toe entirely in white, his bearded face topped with a stylish white Panama hat. Not only was Richards encouraged by the fact that this was undoubtedly the face of the man who had appeared in his dream, but he was also astonished to find that this man seemed to look straight at him.

"You can come out, now, *Dottore* Richards," he said. "It's me. Merisi."

As Richards emerged from his hiding place, Merisi moved towards him, and with a warm, firm handshake, said: "I am delighted to see you again!"

"How did you know I was there?" asked Richards, still a little hesitant and wary.

"I am a very difficult man to hide from, *Dottore*," said Merisi, with a smile. "Come. There is plenty of time to answer all your questions – but, first, I need to get you away from here."

Merisi placed a reassuring hand on Richards' arm, and started leading him away, but the young engineer refused to budge.

"First, tell me what you meant when you said that Grazia is not what she seems. On your word, I've left her alone, in the hospital."

Merisi sighed.

"Dottore," he said, "if I told you that I strongly suspect that your 'Grazia' is actually an advanced, LBC-23 android, capable of shape-shifting and telepathic manipulation, and that this android is designed specifically for the purpose of targeted assassination, would you believe me?"

Richards looked at Merisi in amazement and took half a pace back – but said nothing.

"Of course not," concluded Merisi. "And, of course, you are a man of science!" Merisi reached into the inside pocket of his jacket and produced his disc-shaped communicator. "And what does your 21st Century science make of this?" he asked, at which point Richards was startled by a bright, blue flash, similar to the one that he'd seen just prior to Merisi's appearance – and then the man in the white suit was nowhere to be seen. After no more than a second, Merisi reappeared, after the same burst of blue light, but this time he was several yards away near the main entrance of the hospital.

As Merisi returned to where Richards was standing, the two men looked at each other. No words were spoken. Richards was too shocked to say anything – and Merisi felt sure that he had made his point.

At length, Merisi said: "Look, this is all a lot for you to take in, especially in your induced state of confusion, but you must let me help you."

"Why should I trust you?" demanded Richards, still feeling the need to adopt a defensive position.

"You trust *me*?" exclaimed Merisi. "I think it's more a case of me trusting *you*, with all the things I am showing you, and all the privileged and very dangerous information that I have been sharing with you."

Merisi then reached into his other jacket pocket, and drew

out a small, transparent tube that Richards could see contained a number of small, yellow lozenges. "Take one of these and let it dissolve under your tongue. That will help to make things much clearer for you."

Doubtfully, Richards opened the tube and tipped one of the lozenges into the palm of his hand.

"What is this supposed to do?"

"Well, to borrow an analogy from your own research, *Dottore*, your thought processes are experiencing as much turbulence at the moment as a free-standing wind turbine in the middle of the North Sea in a hurricane! This small tablet will calm the turbulence in your head."

Richards was reassured by the fact that this well-meaning stranger was acquainted with his work – but nevertheless felt that he was taking a massive leap in the dark as he placed the yellow pill into his mouth.

Merisi smiled.

"Follow me – and stay close," he said. Stopping suddenly, he grabbed Richards and stared menacingly into his eyes and whispered, "there are the cryptic images on the wall where I am taking you, for a reason I will explain. Once the drug I have given you takes effect, you are then able to see the hidden symbolic messages in these images which glow and stand out in these images so that only you can understand the messages."

"Walls have ears and if the GIATCOM androids were able to lipread and find out where you intend to look for the missing artefacts in Pompeii before you do, then I am not chancing these androids being invisible and finding their way into the room I am taking you to. Our conversation has been lipread once before, so I am not taking any chances again."

"So I have produced some pictures to represent symbolically what you need to be aware of regarding these androids. I will leading the small party stopped opposite a door numbered also talk cryptically rather than spell things out to you, if anything in particular I tell you is of high importance. For all we know, this current conversation may well be heard or seen by them."

Richards, startled, began to panic and turned his head vigorously as his eyes tried to survey the darkness all around him.

"Relax *Dottore,* like I say, we haven't divulged any information that GIATCOM could use to their advantage in our current conversation so far, so let's just walk and no talk"

Ben Richards remembered very little of his journey with Merisi through the Milanese night. Merisi walked briskly, and heeding the man's instructions, Richards followed close behind, his gaze fixed on Merisi's left shoulder. Occasionally, as he plunged into dark alleyways and hurried up and down short flights of steps, Merisi would glance back and look into his companion's eyes – but said nothing, and his pace never slackened.

As the pair marched on, Richards experienced a number of sudden flashbacks, which grew in clarity and regularity with time. He remembered the sight of the man on the high wire, with the Ponte Vecchio in the background – then he saw the smile on Grazia's face, in the restaurant's candlelight of the Hotel Savoy, as if she were standing right in front of him – then the sound of a passing motorcycle brought to mind the image of the motorcyclist in the wing mirror of the car, as he and Grazia sped along the motorway. Soon, he could piece together other fragments of events from the seminar that came into his mind. At first, they were random images of faces and locations – the young, brash economist, the old man who had introduced himself during the interval, the session chair, the small groups of people who were milling around in the courtyard outside - Boyle and Buttgereit.

It was just as the two men were approaching a brown double-door on the side of a very large brick building that details of the events that took place outside Caravaggio's restaurant came back into Richards' mind with a cold, stark clarity. Immediately, he grabbed the white-sleeved arm in front front of him and pulled it with such force and urgency that Merisi stopped and was swung around to face him.

"Grazia was also taken away by Feng, wasn't she?" yelled Richards, whose mind was partly filled with horror at what might have befallen the young woman, and partly filled with guilt that he had not quizzed Merisi about Grazia's whereabouts if the person who had comforted him in the hospital ward were not here. But his primary concern was to get an answer from Merisi: "If you know where she is, then you must tell me!"

"Welcome back, *Dottore*," replied Merisi, wearing an expression of some relief. "Yes, Grazia was taken away. I have no idea where she is now. In fact, you were both abducted by some very dangerous people. That is why I am so relieved to see you again, as I felt my mission had failed completely – although the fact that you have escaped capture suggests to me you were released deliberately, and that these people want to observe your next moves. That is why I want to get you out of sight as soon as possible. They will be searching for you."

Richards opened his mouth as he began to ask Merisi about his captors, but Merisi silenced him with a raised palm:

"No more questions, *Dottore*. Unfortunately, Signorina Rossini is beyond my reach – for now. Let's go inside."

With that, Merisi turned the round brass knob of one of the doors, and leant against it, as if he knew the door would be unlocked. He beckoned Richards in and then swiftly but quietly closed the door behind them.

The drug that Merisi had given Richards was now having a full effect, and just as had happened just prior to Shui Feng rendering him unconscious outside the restaurant, the young man's senses were suddenly heightened to a degree that he had never before experienced. Even without this sensory enhancement, Richards was immediately aware that the room in which he and Merisi were now standing was part of a nightclub. With the establishment now closed, the building was deserted, but a mirror ball still rotated in the centre of the ceiling throwing faint fragments of light onto every surface, from the "Uscita" sign above the door that they had just used, and from the handful of dim wall-mounted lights that still glowed in the murk.

The room was still heavy with the pungent smell of the crowd that had crammed the room up to just an hour ago – but to his amazement, Richards was aware of the distinct nature of every odour. He sensed the aroma of every type of perfume and cosmetic that had been carried by the night's clientele and could distinguish every stale remnant of the alcoholic drinks that had been consumed there. He detected the scent of perspiration, and the unmistakable vile whiff of marijuana, the kind of smell that used to make him wince at the smell which he had not smelt since his days as an undergraduate student in London.

But the effect of the tablet that Merisi had given him had affected him in a way that was completely beyond his experience. Light, too, was enhanced to the nth degree, and Richards could clearly distinguish the different wavelengths of the shards of light that rebounded off the tiny panes of glass on the mirror ball. Some of the colours he could not name, as he stood open-mouthed, marveling at the gently transforming kaleidoscope around him.

As he watched him gazing around the room, Merisi was well aware of what Richards was experiencing. "It must be a strange feeling for you, *Dottore*," he said as he pulled up two wooden chairs from a nearby table.

"'Strange' doesn't begin to describe it! What on earth was in that tablet you gave me?"

"The precise chemical content is not important. What is important is that it enables you to concentrate on what I am about to tell you, and to think clearly. Your thought processes must be direct, uncluttered and unaffected by any 'noise' caused by events in your past life, your anxieties, your insecurities or your feelings for Signorina Rossini."

Richards found that he was immediately able to concentrate entirely on what Merisi was saying, being aware of other thoughts, emotions and messages that his heightened senses were sending him yet having the instinctive ability to compart mentalise them in such a way that they did not interfere with his main focus.

"Look at me," said Merisi, as he sat Richards down on the chair beside him.

Richards looked into the man's eyes, which had at first seemed to be of a deep brown colour, but which he now could see possessed a dark green inner ring surrounded by sharp triangular flecks of yellow and amber which penetrated the dark brown surrounding. He could also distinguish every pore and every crease of Merisi's tanned skin, and each individual silver-grey hair that flecked Merisi's eyebrows and beard, few though they were, were clearly visible to him.

Merisi's expression became sterner as he leaned toward Richards slightly.

"I need you to focus. Now, do you remember what I said to you in the apartment, back in Milan?"

Immediately, and seemingly without any conscious effort at all, Richards could recall the entire conversation. "Yes. Yes, I can. I can remember everything. Hey, this drug is incredible! Everything that you told me is crystal clear. Come on, now – you must tell me what it is that you have given me."

"It is a drug developed some time ago – well, a few decades into the future, – for military purposes. As I told you, it enables the mind to focus, eliminating the normal 'noise' and turbulence that the brain is prone to. In the days when armies were still largely made up of human beings, the drug was very useful in enabling soldiers to go into battle with heightened senses, and to concentrate entirely on the orders that they had been given, and to not be hindered by their conscience or any latent moral objections they might have to killing others."

"I see what you mean by eliminating the turbulence in the head!" said Ben. "Someone could make a fortune out of something like this. I could market it as the Coanda Drug, with applications for-"

"Yes, I am quite sure that you could use the drug to develop an entire business plan," interrupted Merisi, "but it is very important that you focus on the things that I am about to tell you right now."

"Yes, of course," said Richards, suitably chastised. "I am listening."

"Good. Now, you will remember everything that I told you in the apartment..."

"Yes."

"So, you will understand that there is a great deal at stake – and that is why I have been sent through time to find you and to protect you. This is not something that the authorities will have done lightly!"

"I can certainly remember our entire conversation – but there are some things you told me I find very hard to believe, frankly."

"Believe me, *Dottore*," said Merisi, gravely, "what I told you is very true, and very real. Certainly, the danger that you are in is very real, and could be very close at hand." Richards said nothing, and so Merisi continued: "Now, cast your mind back to what happened to you outside Caravaggio's restaurant."

Again, the events of the kidnapping were instantly crystal clear in Richards' mind, and this was clear to Merisi from the sudden change in the young man's expression. He was now deeply troubled by what he remembered.

"Now," said Merisi, "tell me everything that you remember, from the time you left the restaurant."

Richards described, in great detail, the scene outside Caravaggio's, and of the arrival of the two sports cars. Then he recounted all the information his senses had gathered about Grazia being taken. They were very detailed memories, sharp fragments of images and sounds. Merisi listened, and analysed every word, but there was nothing in Richards' account that he could use to work out where Grazia might now be kept.

For Ben, it was a difficult process since even as he spoke; he, like Merisi, was discovering the facts of what had happened that night for the first time.

After he had described the second car – the one in which Shui Feng carried him away – he suddenly paused.

"What is it, *Dottore*?"

"Well – after that, everything got a bit weird."

"Believe me," said Merisi, with a smile, "nothing you tell me will surprise me, or seem weird to me."

"Well, as I was picking myself up off the ground, this woman got out of the white car and walked towards me."

"A woman?"

"Yes. It was the same woman who had come up on stage during the seminar. I don't know who she was, but... well...I had arranged to meet her that evening. It was because of her that I was in *Caravaggio's* that night." Richards looked at Merisi rather sheepishly, embarrassed at having to admit to being so easily seduced by the woman's dark brown eyes and alluring charm. But Merisi was unconcerned about the detail of how Richards had come to be in the restaurant. He was now far more interested in confirming the identity of this mysterious woman."

"Continue!" he said, dryly.

"Well, she came up to me – and I recognised her, from the seminar, of course ..." Richards again paused as if analysing his own recollections. "Then, I became all sort of... hypersensitive...

like I am now! Was I under the influence of some drug, just then?"

"Yes, that will have been your 'Coanda drug' kicking in, *Dottore.*"

"But how was I given such a drug? Did you-"

"Never mind how I gave it to you. Just concentrate on telling me everything you remember."

Richards did as Merisi asked.

CHAPTER 6

THE FUTURE & AI

Dr Richards leaned closer and locked his eyes on Merisi's. "Well, I told her that Grazia had been taken away, and she told me that Grazia was gone. That there was nothing that could be done about it, and that she was going to look after me. I am sure this woman had something to do with Grazia's disappearance, perhaps?" Merisi sighed as he shook his head.

"Undoubtedly," said Merisi, with an air of resignation.

"What happened then?"

"This is the really weird bit. One moment, I was talking to a beautiful woman – the next moment, there was a man standing in front of me, spraying something in my face. That must have been a hallucination - right?"

"Describe this man!" demanded Merisi, ignoring the question.

As Richards was describing Shui Feng's oriental features and athletic build, Merisi rose to his feet and started to pace across the room. Richards halted in his description.

"So, there can no longer be any doubt," said Merisi, apparently muttering words to himself.

"You know this man?" asked Richards.

"'Man?" exclaimed Merisi, with a rather macabre chuckle. He now stood, facing the wall at the end of the partially lit room, one hand on his hip and the other on his head. He was clearly deeply disturbed. After a few pensive moments, he turned towards Richards, "If it were a man that we were dealing with, then I would not be so worried."

Richards was confused but intrigued at the same time.

"So, who was he? And where did he come from? Are you telling me I was not hallucinating?"

"Oh no, he was – and is – very real! In fact, it was the attractive young lady who was closer to being a hallucination, and it was the drug that I gave you that enabled you to see your would-be assassin as he really is."

"Assassin?" exclaimed Richards.

"Yes! Congratulations, *Dottore*. As I suspected, you came face to face with Shui Feng - and survived!" Merisi sat down in the chair beside Richards, again. "Do you remember a throwaway remark I made about a shape-shifting LBC-23 android?"

Richards nodded.

"Well, it's as I had feared. Shui Feng clearly took on the guise of a woman no doubt he knew you would be attracted to in order to get close enough to you, to kill you! What worries me, though, is the fact that you are still alive. The only reason that you still live and breathe is that Feng and GIATCOM have another purpose for you. Their objective is clearly one of abduction – but we are unclear as to their ultimate aim."

Ben Richards was, by now, becoming increasingly hard to shock, but he was no less anxious to learn more about this android assassin.

"So, tell me more about this Shui Feng."

"He is a highly sophisticated shape-shifting android. His usual modus operandi is to kill, swiftly and clinically – that is why you are lucky to be alive."

Richards retained some scepticism about the detail of what Merisi was telling him.

"So, you are telling me," he began uncertainly, "that this Feng can shapeshift – to appear in any form that he chooses?"

"Well, not any form. But, yes, essentially, he can adopt a very different appearance, and to impersonate another person convincingly enough for you to not know the difference."

"And this impersonation," continued Richards, still requiring a little convincing, "includes being able to give himself an instant gender change, I suppose?"

Merisi looked at Richards with a look of genuine surprise and confusion.

"Gender?" he repeated. "Why would an android need a gender?"

Richards immediately realised that this was a fair enough question.

"Shui Feng is a machine," continued Merisi. "A very advanced one technologically, but he is a machine, nevertheless. That is all part of what makes him such an effective assassin. He has no trace of humanity, no scruples and so there is never any danger that he might hesitate in his mission. He is designed to end the life of his target as quickly and decisively as possible. But Feng has been given different orders this time."

Richards was still struggling to believe that the kindly smile that had greeted him when he had regained consciousness had really been nothing more than a cloak worn by a dangerous assassin.

"But I was sure that it was Grazia that I saw in the hospital," protested Richards.

"Yes, I am not surprised," replied Merisi. "Feng is very good – and he has the advantage you will not begin to suspect he is capable of such deception. I am fairly adept at the art of disguise," he said. "On two occasions now, a grey-haired old man has come to your aid."

Merisi punctuated the suggestion by raising his eyebrows in a meaningful way as he fixed his gaze on Richards' face. As he did so, Richards instantly recalled the old man pouring him a glass of water at the seminar, then turning up unannounced at Caravaggio's restaurant. As he stared back at Merisi, his heightened senses were now able to clearly identify the same dark brown eyes that had previously studied him in earnest from the vantage point of a much older, and far more wrinkled, face.

"I am not a shapeshifter, like Feng," explained Merisi. "With me, it is more a question of artistry, and a little 'theatre craft' – and a little help from some 22nd - century technology which I am not at liberty to divulge, *Dottore*."

The scientist and science fiction aficionado in Richards was fascinated by the glimpses of the future that Merisi was giving him. "It appears that artificial beings play an important part in your world," he prompted, inviting Merisi to tell him more.

"You see *Dottore,* after your death, humans evolved into addictive beings. The global economy degenerated due to people's work being replaced by machines, and that work time being replaced by time spent on addictions. The problem was one of needing to channel bad addictions such as sex, porn, gambling, alcohol and drug abuse, and replacing them with 'addictions' that can help society. A society based on directing our addictions instead towards caring for others, animals and the environment, whilst becoming productive with generating ideas within the future AI workplace. We needed to find ourselves becoming masters of the work life balance in which a 'hyphen economy' had developed, where everyone aspired to have hyphenated careers by being able to carry out multiple careers as an identity. Suddenly, humans became addicted to self-betterment and to gaining different experiences but needed direction."

"This is why your own addiction is so important, *Dottore*-you can give them direction. You see, with all this spare time, humans had the opportunity to develop a multi-career identity, but their bad addictions distracted many from being able to drive their addictions for good intent. You *Dottore* can inspire good addictions, as you are driven by helping save the planet with your talent for innovation. However, to be a breakthrough scientist, you must first be an intuitive creator learning through your own experience... You must first be able to visualise the big picture and develop innovations influenced from your personal experiences, then perfect the detail within them, before you discover what you are looking for."

"Yes, I have always believed that," replied Richards.

"It's good that you have that philosophy *Dottore*, as you are a future exemplar, a future role model. In the not-too-distant future, productivity in the workplace will be controlled by machines, with humans spending their time with their addictions. They need to channel their time towards good intentions, and who better to use as a role model but yourself? They need to apply themselves to the problem of improving the world around them."

"The one thing that machines, or androids, lack is the ability to assimilate information from an environment that has different economic, political, and scientific climates and resources, to create innovative ideas that are inspired by differing factors. Machines can use their artificial intelligence to 'think' for themselves and generate ideas, but they are incapable of anticipating how their ideas may be affected by, or affect, the global situation. So in other words, they can think of an idea and how it affects something, but are unable to think ahead and see how an idea can affect a number of other factors."

"So, androids – like Feng, presumably lack the ability to visualise the bigger picture?" reasoned Richards.

"What they lack, *Dottore*, is 'Multi-collinearity' - the ability to consider how an idea effects a number of variables, and vice versa. To put it another way, it is a phenomenon in which one predictor variable in a multiple regression model can be linearly predicted from the others with a substantial degree of accuracy."

"Yes, I understand that," said Richards. "In my own field, if a stadium is designed with the structural engineering, which supports a roof-mounted turbine array of sufficient size, then that machinery can generate enough power to act as a centralised power station for local buildings. If you adjust the structural tolerances, you can adjust the size of the turbine array that can be structurally supported, and so change the amount of power generated. Similarly, if you adjust the curvature of the stadium roof to reduce the incline, less attached windward flow will travel up the curve to produce a more constant flow, resulting in less output being generated. The two independent variables – the structural engineering and the architecture's aerodynamics - will affect the dependent variable of energy generation."

"And that is something that androids are incapable of achieving," added Merisi. "To be able to grasp such concepts, androids will need to be fitted with some sort of prefrontal vortex in their cybernetic brain. So that they can be prewired to produce different ideas that affect a design being complete, and so produce innovative ideas from abstract reasoning, like humans do."

Richards nodded in agreement.

"Your world will continue to be driven by a global capitalist model, which implies dependence on ideas being generated from fluid intelligence. A greater need for abstract reasoning from humans to develop their fluid intelligence is required to make best use of our resources for the robotic age. With cybernetic androids being incapable of providing innovative ideas to best suit a given situation, and being unable to consider different variables intuitively, humans will continue to play an instrumental leading role in the future. Androids act within the confines of parametric thinking and are goal oriented to complete a task for which they have been programmed - a deterministic logic that they follow as part of a system. Humans will be indispensable due to their ability to create unique ideas, and to create the correct philosophical system for machines to follow. They do that by deriving innovative ideas from sourcing, and by assimilating information from different variables. A human can obtain information from a given environment at any given time and react independently within a changing situation."

"To gain investment in the future, there is a need for the correct assessment of information from different variables. It is also important to make the best decisions in a given situation to ensure stability, growth, and a renewal process that is credible and sustainable, and has longevity."

"Understand, *Dottore,* that information and ideas became the future currency of trade and exchange within the global economy after the middle part of the 21st century. They replaced money in transactions. Also, at this time cybernetics, instead of being used for the advancement of industries such as manufacturing, began to be misused, resulting in the destruction of our planet towards the end of that century. What the world really needs is a humanoid with cybernetic brain implants, who is able to steer cybernetics in the right direction. Imagine such a system integrated in the brain of an astronaut where deterministic logic and an aptitude for solving parametric tasks are required in high-pressure situations, and where high concentration levels are of the utmost importance. Reducing the risk of human error due to poor productivity and memory loss is critical in these situations. The future humanoid astronaut will, of course, be able to use human creative thought processes to be able to adapt to changing environments in space and on other planets. By making use of cybernetic implants, humans can become more focused, and be free of the distractions of cravings from potentially bad addictions. No longer being fully human also minimises the risks posed by animal instincts, urges and frailties that cause bad addictions. Conversely, the future human can be creative due to the natural ability to be productive in generating ideas. The future humanoid will become the master of compartmentalising multi-occupations within the hyphen economy of the future. Sustaining multi occupations will be the norm in the future and will be the key to achieving the desired work-life balance."

Merisi paused, realising that Richards was beginning to experience information overload. He rose from his chair and slowly walked across to the far wall whilst running his right hand through his shock of jet-black hair. His left hand rubbing back and forth across his bearded face, "But let us return to the problem of Feng," he sighed.

Reluctantly at first, Merisi told Richards everything else he knew of his old foe and did his best to satisfy the young academic's curiosity as to the science behind the Artificial Intelligence Programme that had developed such an efficient and dedicated killing machine. He was careful to avoid transgressing the strict UNA protocols that prevented him from disseminating knowledge of scientific discoveries that were not to be made for several decades. What was most important at that moment in time was to warn Richards of the danger that he now faced.

"What you need to realise, *Dottore*," he said, returning to the point, "is Shui Feng might be anyone. When he appeared to you as an attractive local woman, he might have been impersonating someone that he met in town, or equally it might have been a persona of his own making. What I am now sure about is that the person you thought was Grazia, in the hospital, was actually Feng."

Richards was suddenly stricken with grief and anguish at the apparent loss of his young companion. It briefly occurred to him that his emotions might have been exaggerated by being under the influence of Merisi's drug, but his sense of loss was no less real. He had never felt pain like this before. The death of his mother had affected him all his life, but this was something different. This was a tangible, visceral feeling of emptiness and longing. The stress of separation was heightened all the more by the helplessness he felt in being unable to help Grazia.

In desperation, he turned to Merisi, and said: "Is there nothing we can do? Can we not at least report her to the authorities as being missing?"

"Not a good idea, *Dottore*. If the local Police become involved, and they happen to get in Feng's way, it could become very messy! Please understand interfering with history is not something ever done lightly. For us, it is the very last resort; a final solution, when all seems to be lost, and when there is no other solution available. My mission is – or was – to prevent your assassination, in an effort to right some of the wrongs that were perpetrated some 50 years from now." Merisi looked away and stared pensively into the gloom on the far side of the room. "The fact that I so easily managed to succeed in my mission somewhat troubles me."

"Well, I for one am delighted about that," exclaimed Richards.

"Yes, yes, but I had been working on the assumption that Shui Feng was sent to follow me, with the intention of ensuring that you died at the seminar – but we now have intelligence that GIATCOM has some other purpose in mind for you. Feng has had every opportunity to end your life. I'll be honest with you – when I left the restaurant that night, I fully expected to see you dead in a pool of blood. That is very much his style."

The blood drained from Ben Richards' face; but this was not because of the realisation that he had been close to death himself, but because of what might be happening to Grazia. "And this is the person, or thing, you think is holding Grazia?" he demanded.

"Yes," replied Merisi, "but the good news is I didn't see Grazia's body out on the pavement either, so my guess is she is still alive. Again, I do not know how your friend will fit into Feng's plans – but he must have something in mind for her as well. One thing is certain – Feng did not spare either of you out of sentiment, or to keep the piazza tidy!"

"You are guessing that Grazia is alive?" said Richards,

becoming even more perplexed.

"Well, that is all I can do, for now," replied Merisi. "Nothing is certain. From the moment that you walked off that stage in the conference hall alive, every nanosecond of reality has been part of a new future being created. One which we hope will be better than the one I left behind. Already, the world in 2112 that sent me here might be one I do not recognise. Once an event is changed, there is no way of knowing what the impacts will be – as I am sure you can imagine, *Dottore*."

"Yes, of course," replied Richards with an air of confidence. "It's chaos theory – which says that the air displacement caused by one butterfly flapping its wings in, say, Brazil can affect weather systems in the United States, maybe even altering the course of a hurricane."

"-or causing it or preventing it! Yes, in theory. The theory is correct – but how many butterflies live in Brazil?" It was a rhetorical question, and so Merisi did not wait for an answer. "A countless number, no doubt," he continued, "and I would say that, in practice, the wing movement of virtually every individual butterfly will have a negligible impact on its immediate surroundings, an impact that soon dissipates to zero."

"*Dottore* – you are a genius."

Richards raised both eyebrows, and opened his mouth, as he was about to utter some banal statement in order to appear modest – but Merisi cut him short, by continuing: "That is not a huge compliment, as there have been many thousands of people who have lived who could have been described as having a certain 'genius.'" "Some, of course, have made their mark in the world, and have become very famous, but far more of them have either squandered their talents, or wasted them on trivial pursuits. In your case, *Dottore*, my fear is that you will miss the opportunity to fulfil your potential. It would be very easy for you to now just go away and live an unremarkable life. You could become a lecturer on wind and fluid dynamics, no doubt, and live in a nice house – you could, in effect, be just another butterfly beating its wings, but with no lasting or tangible effect – but we believe that you have far more to offer than that."

"So, why me? How can you pin so much hope on me?"

"That's a very good question. I admit that we had very little to go on. After all, until very recently, you died at the tragically young age of 33 – but you were right! You are right! All the technical information you need is already available in your early papers. After your death, your work was totally ignored – unsurprisingly, when you think about it, as nobody had a vested interest in following it through. But I am going to make sure that it will all be different this time. Obviously, I need to keep you alive, and also provide you with all the support that you are going to need in order to nurture your work to fruition."

Richards paused to think about what Merisi was telling him, and it was now his turn to rise from his seat and pace across to the far wall. He turned and stared at Merisi, studying the man's dress, his shock of jet-black hair and the tanned, bearded face looking back at him, sternly and silently.

"If everything you are telling me is true," said Richards at last, "and I don't think that I have any choice but to believe you, then how do you think you are going to help me? Are you going to sponsor me?"

"I can do better than that, I can give you hope and belief," as Merisi stood and walked slowly to Richards, placing his right hand gently on Richards' left shoulder while staring into his eyes. "You see most people in your present world want to be labelled with letters after their name, for their own personal narcissistic gain, others lead through their own experience and natural talent."

"You are a rare breed, Dr Richards. Some people create their own destiny, and for some destiny chooses them. Greatness can be created or thrust upon us. Some people achieve greatness in one or many things in their lifetime. Others may try and never will. If you are genuine, you will achieve great things *Dottore*. Shedding your old self *Dottore* will allow you to feel free to experience an inner rebirth. Leaving everything behind to enter the unknown to encounter countless unconscious monsters, you will finally return home with a sense of renewed fulfilment and wisdom."

"Your ideas cannot be developed without taking your spirit, and your spirit cannot be taken. Ideas are futile without your passion and belief coming from your spirit, to develop your ideas. So your ideas will not fully succeed in someone else's mind. Let your spirit ascend from the evil that surrounds you to being renewed through your transformation and newfound vision to become a reality."

"Out of suffering you will emerge stronger in character labelled with scars rather than letters, on your journey of wisdom. We are all born with souls *Dottore*, but not all of us know how to fully embody and integrate our souls into our human experience. May your continued journey spiritually awake continued unique ideas and beliefs."

"So, you mean I will be walking alone on my journey?" Richards asked.

"Precisely, *Dottore!*"

"Spiritual ascension also known as spiritual awakening is a natural evolutionary process which involves the process of shedding the old self and experiencing an inner rebirth. Out of suffering have emerged the strongest characters, from scars on their souls."

"The truth is that in our modern world, we live ego-centrically rather than soul centrically. As Saint John wrote a philosophy - 'The Dark Night of the Soul':

If a man wishes to be sure of the road he's travelling on,
Then he must close his eyes and travel in the dark.

Traditionally the Dark Night of the Soul refers to the experience of losing touch with God/Creator and being plunged into the abyss of Godless emptiness.

At some point, most of us go through a phenomenon known as the Dark Night of the Soul.

Although we try to run from it, it is still there.

While some of us seek reprieve in religious thoughts, others of us seek respite in spiritual philosophy or psychology, and still others seek relief through addiction and mind-numbing external pursuits."

"The truth is that although we are all born with souls, not all of us know how to fully embody and integrate our souls into our human experience and good addictive pursuits which are healthy for the body and mind, such as exercise and creativity."

A sense of pride gradually filled Richards' face as his frown transformed to a raised eyebrow and a nod of his head, recognising his mission had been chosen for him, and people were relying upon him, "But why must I go to see the future?"

"Because you must realise your full potential of creativity, which will be stimulated by seeing what has happened to the future without your visions and ideas being created," says Merisi.

Richards looked in disbelief.

"Your creative potential is innate and has been developed through your personal experience so far in life, as by nature you have a form of autism that allows you to look at situations in a more divergent thinking way, but also long-term Post Traumatic Stress Disorder has given you that creative drive stemming from the tragedy you witnessed of seeing your mother die in a Tsunami caused by climate change. You have the ability to use your unique creative talent in a self-motivated way driven by your PTSD to succeed, but you must have a more focused purpose to see with your own eyes the devastation and the number of natural disasters occurring more in the future due to the absence of your creative ingenuity and visionary leadership. Creative leadership can't be taught you have to develop it further through your own empiricism."

"Shades of darkness will come to the foreground.

The world will be in darkness if we cannot supply enough power to meet demand. Darkness will fall on all of us, if we rely rely too much on renewable technologies making use of sunlight and energy being produced from heavier shades of operation, such as nuclear power generation creating a dangerous operation and waste disposal, besides its high cost and danger from human error and terror attacks, leading to Armageddon."

"Some people choose to be a leader of their own destiny and some a leader of the planet's destiny, whilst others are chosen to be a leader of their own destiny that leads the future of the planet's destiny! The latter has already been chosen for you. Others have already chosen you as their leader. Wear the honour with pride, Doctor. You must figure out what the SATOR SQUARE words personally mean to you on your journey, to help create your design philosophies. That is what is magical about the square; it helps crystalise conceptual thinking tailored to the chosen one. That chosen one is you. It was not only used as a code for time travel by spiritualists and the church, but it is a code that suggests the path to follow to achieve utopia for the planet. The chosen one not only has to use the SATOR SQUARE to believe in and experience time travel, but understand what the SATOR SQUARE code personally means to them, to be able to create a path for us to follow. It is down to you to crack your own personal code, in relation to the SATOR SQUARE's code. The future of the planet depends upon it."

"There will be a word coded by UNA, you will see written on a wall when you go to the future. It is cryptic as it is protected from GIATCOM figuring out what UNA is trying to tell you and is associated with what you have to carry out. It is coded so that you need to work out what it is trying to tell you. As UNA will not guide or assist you in the future, you are to make your own journey and find your own way, as you are to find your own personal inspiration and avoid the risk of any GIATCOM androids getting close to UNA if they happen to follow you. It is all part of resetting a new world here in the present, for you to work it out for yourself and not be told it, in fear of GIATCOM androids seeing, hearing or understanding your intentions, and then intercepting and preventing you from carrying out your intentions."

Richards snapped out of a trance he found himself in, as he listened attentively.

"Although, one thing that I am going to do," said Merisi, as he took his hand away from Richards' shoulder and reached for his waistcoat pocket, "is to give you these." He held out, in the palm of his hand, a transparent, cylindrical tube containing what Richards estimated to be about two dozen yellow 'Coanda' tablets. "These will improve your chances of staying safe, as they will enable you, maybe, to see through Shui Feng's disguises. It sounds like you were able to see Feng as he really is, outside the restaurant – and so it will serve you well to remember that face! The effects will last for three to four hours – but don't, under any circumstances, take more than three in any 24 hour period, as you will probably not enjoy the side-effects."

As Merisi turned to walk away, but then turned back and shook Dr Richards' hand, as his eyes were locked on Richards' eyes once more, "Believe and let your spirit take control. You will then reach your pinnacle of creative thinking with spiritual belief. Hard work, mental focus, and good physicality is not enough to achieve greatness, you must have spirituality to take you further."

CHAPTER 7

THE CO-ANDA 19 VACCINE

"**S**o, what exactly are these drugs, and what do they do?"
"In 2019 there was a pandemic named Coronavirus. It killed millions of humans. Many vaccines were created. The 2019 pandemic was the dawn of a new era in the world's labour market. A new generation of science and technology in medicine was in a high growth period. Biotechnology and advancements in research and development within medicine were the new predominant labour market. So, a cynic would say new viruses had to be manmade so that anti-viruses could be created to serve new and stable economic growth in the next generation's labour market. As the educated youth were predominantly choosing science-based careers in medical research, there had to be a demand for jobs in research and developing more advances in vaccines and medicine to meet the increasing supply of medical research students. Some might use the term 'wolves in sheep's clothing' to describe the money making, but there is **Corruption** and **Greed** in every industry. They are both attractive and addictive as a basic human behaviour well practiced throughout

history.

Richards gazed into Merisi's eyes with disbelief, as he thought this didn't sound like the usual conspiracy theory, it sounded credible.

"A drug called the 'Co-anda 19 vaccine' was created many years after. It was not only deemed to be the best to fend off the virus but also to rejuvenate the senses as well. A higher dosage enhanced sensory perception, whilst primarily protecting vital organs from the attack of deadly viruses."

"So why is it named Co-anda 19?"

"The name was inspired from your wind research you adopted from Henri Coanda's - Coanda effect. Your application of the Coanda effect being used to enhance the performance of wind generation in an urban setting by roof mounted turbine array was seen to be revolutionary with its design philosophy. The premise of your research was to create a scientific design that redirects the predominant wind flow and creates a constant undisturbed wind flow. Your research was never developed, of course, because you were assassinated."

"It is named after the Coanda Effect, because it's the principle of the drug attaching itself to a virus or drug to act as an antidote to it. Whilst depending upon the dosage, it can have a psychological effect of creating a constant flow of lucid dreams in reality to induce imagination, and fast euphoric thoughts. It enhances superhuman abilities such as fluidness of ingenious ideas, whilst killing the virus or the effects of another drug. The higher dosage of the Co-anda 19 drug has been specifically designed for the future 'Chosen One' in mind, so that they can use the drug to decipher the disguise of GIATCOM, and to help promote their ingenious ideas. Its development started in 2024 by scientists inspired by the vaccine drugs being developed for COVID-19. It was to be used against viruses which were variants of the COVID-19 pandemic outbreak, post COVID-19. But since then, it has been adapted to a more potent version. It has been frozen and withheld being used until deemed to use on the right individual at the right time. In the future, scientists identified that individual as being you Doctor, and the right time for you to use the Co-anda 19 vaccine is now in your fight against the antagonists trying to drug you with a virus in order for them to take your ideas.

Richards was very happy to be given the drug, and so he tucked the tube away in his jacket pocket.

"And what about helping me to succeed in my work?" he asked.

"Well, for that, I am going to give you something that will sustain you for the rest of your life as opposed to just a few hours!"

Richards frowned, his expression betraying both confusion and scepticism.

"Dottore Richards," declared Merisi, placing a reassuring hand on Ben's shoulder, "I am going to give you purpose. And I am going to do that by showing you the future!"

CHAPTER 8

CARAVAGGIO'S LOST WISDOM

Merisi led Richards back to where they had been sitting. Both men sat down, and Merisi waited patiently for a moment until the enormity of what he had told the Richards had sunk in. Richards sat back in the chair, his legs stretched out in front of him, and his hands resting on his thighs. "So, you're going to show me the future, are you?" he asked, at last.

"Correct."

"And you're going to send me there?"

"Yes. There are certain things that I want to show you. Do you remember what I told you back in your friend's apartment?"

"Yes, of course," replied Richards, still feeling the effects of Merisi's drug, allowing him to remember every fascinating detail of Merisi's account of the discovery of how time could be bent and manipulated. "So, presumably you know the whereabouts of a gateway, or a portal of some sort?"

"Exactly, Dottore. You are a quick learner," said Merisi, with a smile. "There is one not far from here. I know the precise location and time it is due to be opened by my organisation – and we must not miss that opportunity. As I have already explained, our activities here have already changed history to some degree, but I hope and expect the wings of our butterfly won't have caused too many ripples in the future – not yet, anyway."

"But how can you be so sure that this Shui Feng will not have done something to cause a major change in the course of history? If he's a ruthless and reckless killer, as you say he is, he could do all sorts of damage – couldn't he?"

"No, I didn't say that he is reckless," said Merisi, raising an index finger in admonishment. "He is ruthless, yes, but he is driven by a cold logic. Like me, he will have been trained to operate with surgical precision. The difference, however, between him and me is his lack of humanity. He has no emotional investment in the world, so why would he care if his actions caused a major anomaly in history? Why would he care if his actions were even to cause the very extinction of humanity? He is a precise and focused 'thinking machine' – but he has no interest in any wider consequences of his actions."

Merisi paused. He could see, from the change in Ben's expression, that the mere mention of Grazia being his friend, had reminded the young man of his sense of loss. Merisi was, himself, reminded of the callous and unhesitating way in which Shui Feng had murdered the antiques shop owner and his wife back in Siena. He wondered how much historical damage might have been caused already by Feng's actions, which could undoubtedly be seen as being reckless – but he kept these thoughts to himself.

After a while, Richards' thoughts turned away from Grazia and he became engrossed, once again, in the main matter at hand – Merisi's declared intention to transport him to some future time and place, as part of a grand strategy to change the world's fate!

"So," he began, "the big plan you and your organisation have is to make a major change to the course of history by replacing the future dominance of nuclear power with a sustainable alternative?"

"It is the only plan, Dottore," replied Merisi, with a shrug.

"We are calling it 'Operation Reset.' It is a dangerous strategy, for all sorts of reasons, but it is the only option left to us–believe me."

"Well, if I – we – succeed, then isn't there a danger that you will never get 'back home'? What if the great discovery of time travel is not repeated when the timeline is reset?"

Merisi was already nodding knowingly. "Yes," he said, "that is a distinct possibility. In fact, there is now a risk that Shui Feng and I will be marooned here in this time and space. In my heart of hearts, however, I believe that all the landmark discoveries will be made sooner or later. Men and women will always watch birds in flight and wish to take to the skies themselves. They will always gaze across the sea and up to the heavens and wonder what lies beyond. Humankind will always find a way to fly, and to travel in time, and to escape the confines of the earth. It is merely a matter of time..."

"The problem that we have here and now Dottore, is a portal that connects 2112 with 2017 can only be opened from the 2112 side. That is why we took the precaution of scheduling the opening of portals in and around the location of your assassination at around about this time. My first plan was to provide you with the knowledge and motivation that you require, and then to return to my time and remain there – but the presence of Shui Feng here in 2017 is a worry, and I will need to do something about that."

Richards' imagination was now racing, thinking of the implications of what he might achieve during his life. "So, what will happen to you – all of you – in the future should I succeed in helping to create a world based on more sustainable energy? Will people and places suddenly change or disappear? What if you, for instance, are never actually born in this 'reset' world?"

"These are all very good questions," Merisi smiled. "Quite simply, nobody today or in the future knows the answers. It is true that we have been manipulating time in order to travel back in history, but nothing has ever been done to make wholesale changes like you, I hope, are about to make. That is why nobody knows for sure what will happen."

"What are your expectations?"

"Well, I subscribe to the theory that reality is built nanosecond by nanosecond, and molecule by molecule as time proceeds. That means it is not possible to actually change events once they have happened or reverse the creation of an object or person once it, he, or she has been created. All we succeed in doing when we go back in time is create a fresh reality, a new version of the universe, without destroying the original."

"If that's the case, how many different versions of the universe are there?" asked Richards.

"There are many – and the same number of versions of what we, and you, call 'time'."

"So, there is the very real possibility," suggested Richards, "you will be stuck with the reality you 'currently' have in 2112, and you can merely create a better version of reality for others?"

"*Ecco!*" cried Merisi, delighted by Richards' grasp of the situation. "That's exactly right – but, if that's the case, we will at least have one major consolation; you will understand how important it is to create a better future for our children?"

"Of course," Richards gasped, as his face became stunned with amazement after realising what Merisi was asking him to do.

"Well, maybe it's even sweeter for us to create a better past for our ancestors!"

The two men chuckled together at the absurdity of the situation they were discussing – yet they shared an awareness of the gravity of the moment.

"Nothing is certain, Dottore," said Merisi. "All we can do is the best we can, in whatever time and place that we find ourselves."

Richards felt he could not disagree with the sentiment that Merisi expressed, and so he did not reply.

"So there is a portal nearby, you say?" Richards said, breaking the moment's silence. "And, no doubt, it will be marked with a SATOR SQUARE?"

Merisi was slightly taken aback by Richards' level of knowledge, but was nevertheless pleased.

"Yes, that is precisely right, Dottore. What do you know of SATOR SQUARES?"

Richards told Merisi of his long-standing relationship with Father Luigi, and of the conversation he had had with him in the cathedral. "... But Father Luigi never mentioned anything about time travel."

"I would not expect him to know about that, Dottore," said Merisi, with a knowing smile that was becoming very familiar and rather reassuring for Ben Richards. "It is something known to a select few among early Christians in Europe – and by a similar number of holy men elsewhere in the world - and there are even fewer who are now acquainted with the true purpose of the SATOR SQUARES."

"Actually," said Richards, rummaging around in his inside jacket pocket, "an old man in the auditorium gave me a card bearing a SATOR SQUARE. He didn't tell me his name. Do you know who he might be?"

The two men smiled, now that the secret of the old man's identity was now shared knowledge. Richards gave the card to Merisi, who took it, and gazed at the SATOR SQUARE.

"And on the other side of the card," continued Richards, "there is a painting – by Caravaggio, I believe."

"Yes, it is one of Caravaggio's masterpieces, and there is a very good reason for me giving you this card – and it brings me nicely to what I need to tell you this evening."

Richards was confused.

"Take a look at these drawings," said Merisi, as he produced a device from his pocket, and with what appeared to be a casual wave of his hand, he extinguished the dim light sources being captured and reflected by the mirror ball. He replaced them with a set of images. Each image took the form of a sketch, yet the multi-coloured etched lines appeared to Richards to have a neon quality, and an inner light source of their own.

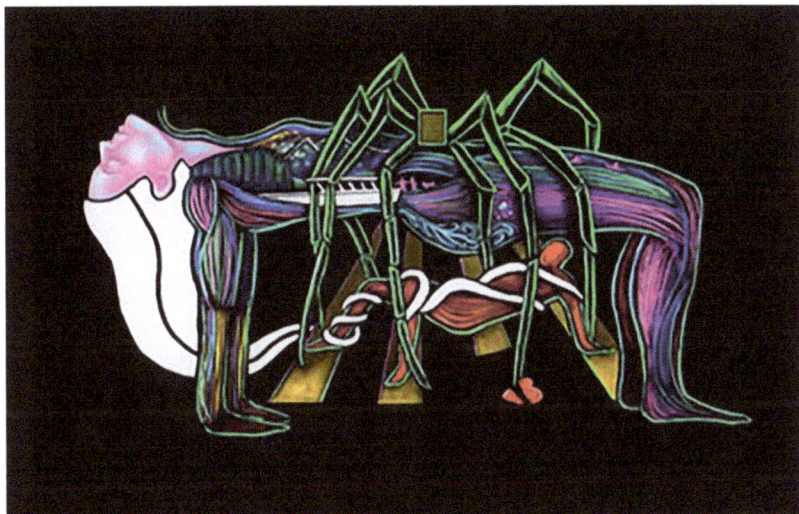

"Sketches," replied Merisi, with deliberate understatement. "They were a final gift from the great man, himself. He illustrated his philosophical ideas on a few scraps of canvas as we sat in his parlour. We chatted about life and destiny, about the past and the future. I say we – but it would be more accurate to say that he chatted while I listened. I updated *Caravaggio's* drawings, so they are relevant to modern times, and they have provided me with guidance in warning you of the dangers that surround you, Dottore.

"It was our final meeting before I had to return to my world, and to my time. There was no way that he could have known, of course – yet he seemed to be imparting a treasure trove of wisdom to me, as a man might do on his deathbed. I very nearly left the drawings in his home, but he insisted I take them away with me. And here are my adaptations," he said while sweeping his right hand in front of the sketches that drew Richards' attention to all five of the sketches. "Of course, the images you see have been enhanced in the preservation process – and I am sure that the level of consciousness you are experiencing at the moment enables you to perceive a wide range of colours in these drawings..."

"'A wide range of colours' would be an understatement," muttered a wide-eyed Ben Richards.

"...but Caravaggio drew these for me on bits of canvas, using a piece of charcoal that just happened to be lying on the table."

"Do you still have the original copies?"

Merisi was immediately in tune with Richards' thought process.

"The 'Caravaggio originals,' you mean?" he asked.

"Yes."

"Yes, I have kept them – but one small consequence of being able to travel through time is that the antiques industry was one of the first to decline!"

Richards didn't dwell on this unexpected implication for long – for he was keen to discover the deeper meaning of the extraordinary images Merisi was presenting to him.

"In the future Dottore, art is used by humans to communicate between themselves, to avoid androids understanding the hidden messages in the art. Emotions created from the Art are only comprehended and communicated by humans. Androids are unable to comprehend their own thought processes through emotions they receive from the art they see. Only humans can do this, and it is their only way of secretly communicating without the androids understanding. Androids can understand human communication if it is written or verbally communicated, but are unable to decode communication through abstract images. A human is able to distinguish whether a being is a human or android by testing their response to emotional abstract images. Androids will become adept at intercepting other communications channels and deciphering traditional passwords and codes."

"So... What is this one supposed to mean?" Richards asked, pointing to a drawing that appeared to depict the ambiguity of a man and a woman representing waves crashing into the face of a cliff, and in the background the sea on the horizon.

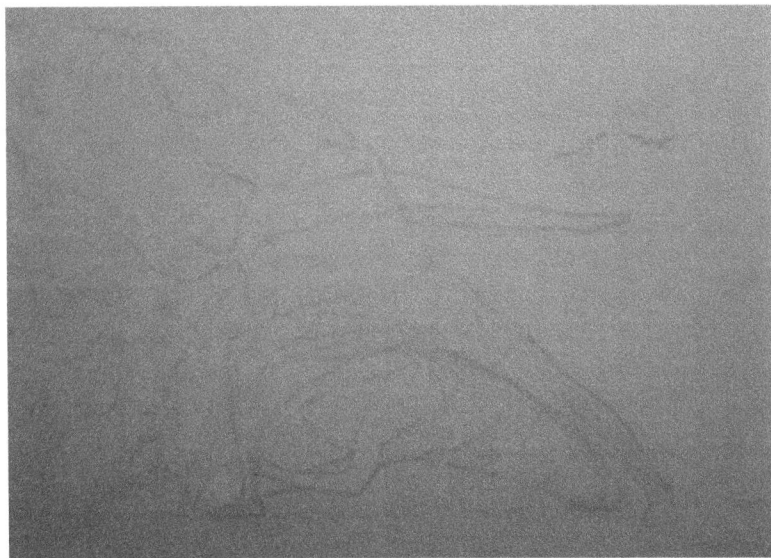

"This is what the great man referred to as 'The Cliff Dancers,'" replied Merisi. "It is a strange coincidence this is the first sketch I asked him about as well." Richards was by now beginning to feel a real connection with Caravaggio through Merisi.

"And what did he say?" asked Richards.

"'Enzo,' he said – as that is what I called myself at that time.

'Just tell me exactly what it is you can see.' 'It is a man and a woman dancing,' I said. 'Look closer,' he said, and slapped me hard, on the cheek. I stared very hard at the drawing, and after some time, I could see the two figures appeared to be conjoined. Sometimes, I could see two distinct figures – a man and a woman – and then in a split second, the figure became one entity with two faces."

Richards peered at the image, and, sure enough, he was at last, able to detect the *trompe l'oeil* - visual illusion.

"Yes, I see it," he said. "Incredible!"

"And Caravaggio did this with a piece of charcoal, Dottore, as if doodling for the amusement of a child. Genius! Sheer genius!" The two men stared at the image with wonder, before Merisi continued with his account of his conversation with Caravaggio: "'What I am trying to tell you, Enzo,' he said, 'is that things, particularly people, are not what they seem, and that it is often great beauty that is the cloak for the dagger.' He went on to tell me that beauty might take the form of a dance, or a painting, or maybe a pretty face. Of course, I was familiar with a world in which shapeshifting and various other highly elaborate and high-tech methods of disguise were common, and so I was already armed with a well-developed sense of distrust of anything that spoke! Caravaggio had no knowledge of such things, but his wisdom still had a profound effect on me, and on how I thought."

Ben Richards was already able to identify with how easy it was to be beguiled by a pretty face. "I could have done with that advice when Feng approached me at the seminar," he said.

Merisi smiled broadly, throwing the laughter lines in his tanned face into more vivid relief. "It appears that Michelangelo Caravaggio is also speaking to you, now, Dottore," he said. "It is so very easy for us to be caught off guard by a thing of beauty – or merely by a kindly old man!"

Richards made no connection with Merisi's remark about an 'old man,' as he was already eagerly studying the other images that hung, magically, it seemed, in the still, musty air of the nightclub.

"What does this drawing mean?" he asked, nodding towards a drawing of a male and female form, where the female appeared to be in a bridge-like pose.

"Ah! That's a strange one. He referred to that one as *'Monogamia'* – the Italian word for 'monogamy.' He laid great emphasis on this, telling me how important it was to draw strength from the love of one's life. It illustrates the need for us all to stay faithful to our love – the one who can help us find our true path and destiny. We all need a soul mate, for companionship, on the journey to the future, to offset our weaknesses with their strengths. Actually, I have always felt that the fact that he was never married himself made his insistence on this point strange."

"Well, maybe there was someone special to him whom he lost?" suggested Richards.

"He never revealed such depths of his own mind," snapped Merisi, "and it would not have been advisable to have pried into such things, as he was as accomplished with his fists as he was with a brush – and very quick to use them."

"What's with the story inside the woman's body?"

"Ah, I am glad you noticed that. Caravaggio is predicting that, in the future, we need to be careful with how we treat our environment. I am sure that this will inspire your own ideas, Dottore!"

"Really? Please elaborate."

"Well, he is suggesting that we view our environment in

much the same way as we view the human body. In much the same way as the human body has its skeletal, cardiovascular and nervous systems, cities have infrastructure systems, such as highways, water distribution systems and communications systems that allow people to move about the city, deliver goods and services and share information. But, as you know, these city infrastructures often collectively fail as a result of man-made and natural disasters. Unlike the human body, these systems are, typically, independent of one another, and are not interconnected with other utilities, nor are they built for adaptability. As a result, there is no guarantee - often no real plan - that they will work well together when subjected to a shock, like flooding, a large earthquake or an extreme heat wave - or when they have had to deal with the long-term stresses caused by population growth and the ensuing congestion."

"The interdependency of these systems can make a city brittle, so that, in a disaster, a failure in one system quickly cascades to the next. For example, in the present time, metropolitan water is used for drinking, but also for local agriculture and power generation. A drought that severely restricts water means that water for human consumption must become a priority, leading to a loss of water for agriculture and power generation. To combat these knock-on effects, your cities must not think about water systems without also thinking about food supply and power systems. As with the human body, each of these systems is semi-autonomous, to capitalise on its specialised function, but it will fail – now or in the future – if it can't rely on the other systems. Hold your breath for a minute Dottore and think through all the system cascades your body will go through – aerobic, circulatory, muscular, nervous system – to adapt to the loss of oxygen. The difference is that your body has evolved with the ability to adapt; many cities have not."

"I can give you several examples from your own time Dottore, where problems have cascaded from one infrastructure system to another as a result of a major disruption. You will be aware of the Fukushima earthquake and resulting tsunami in 2011, when flooding of backup generators caused an overheating of fuel and a subsequent nuclear release."

Richards nodded slowly. The word "tsunami" had always troubled him greatly, ever since the tragic event that robbed him of his mother.

"Similarly with Hurricane Katrina," continued Merisi. "Once the levees broke, the massive flooding disrupted systems relating to emergency response, medical care, water and wastewater systems and transportation services, and eventually caused considerable loss of human life.

"The 'bottom line' is that, from a resilient-city policy point of view, it's not enough for water, electricity or fuel systems to be resilient to shocks on their own. The water, electricity and fuel systems together need to be resilient to these shocks."

"So, what can cities do?"

"Design or redesign. Cities need to design – mostly redesign – their systems for resilience, that is, so that they can continue to deliver services in the event of a shock or acute stress."

"What do you think of these?" Merisi asked, gesturing to some more images with the palm of his hand. He was drawing Richards' attention in particular to a sketch of two identical figures, sitting back-to-back.

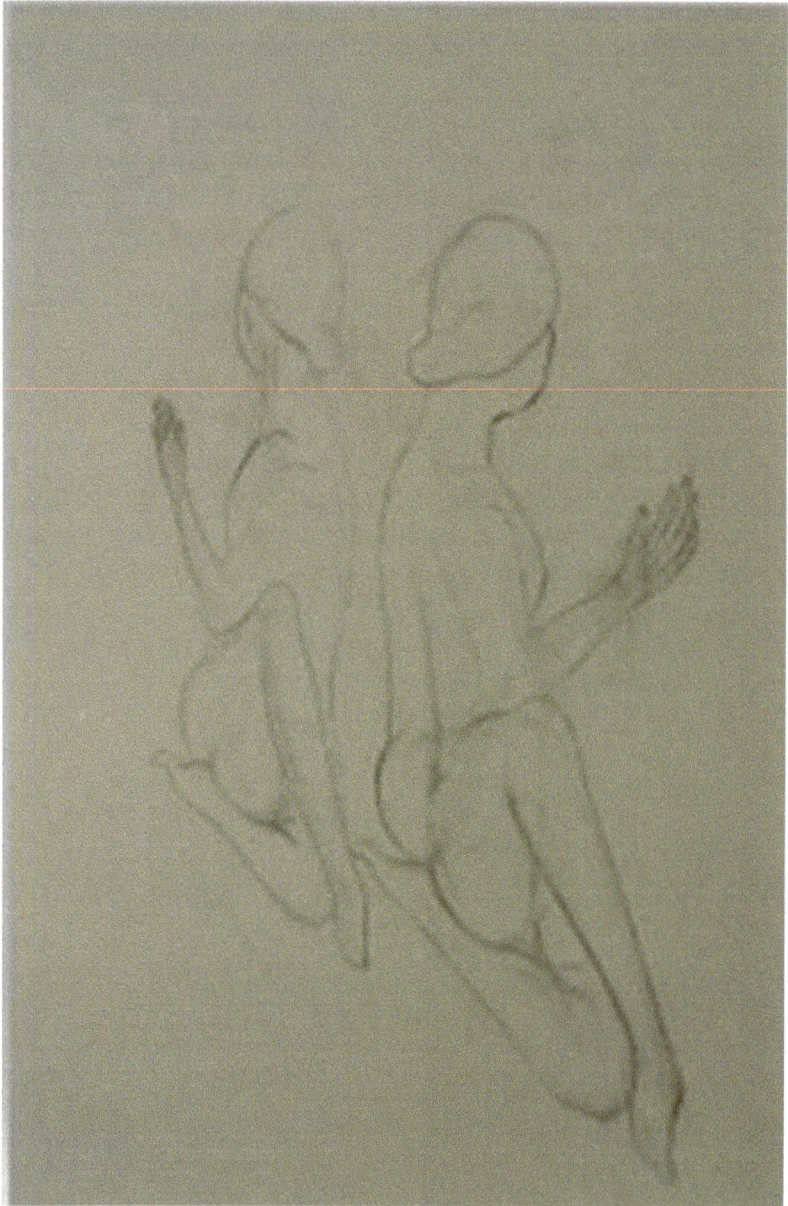

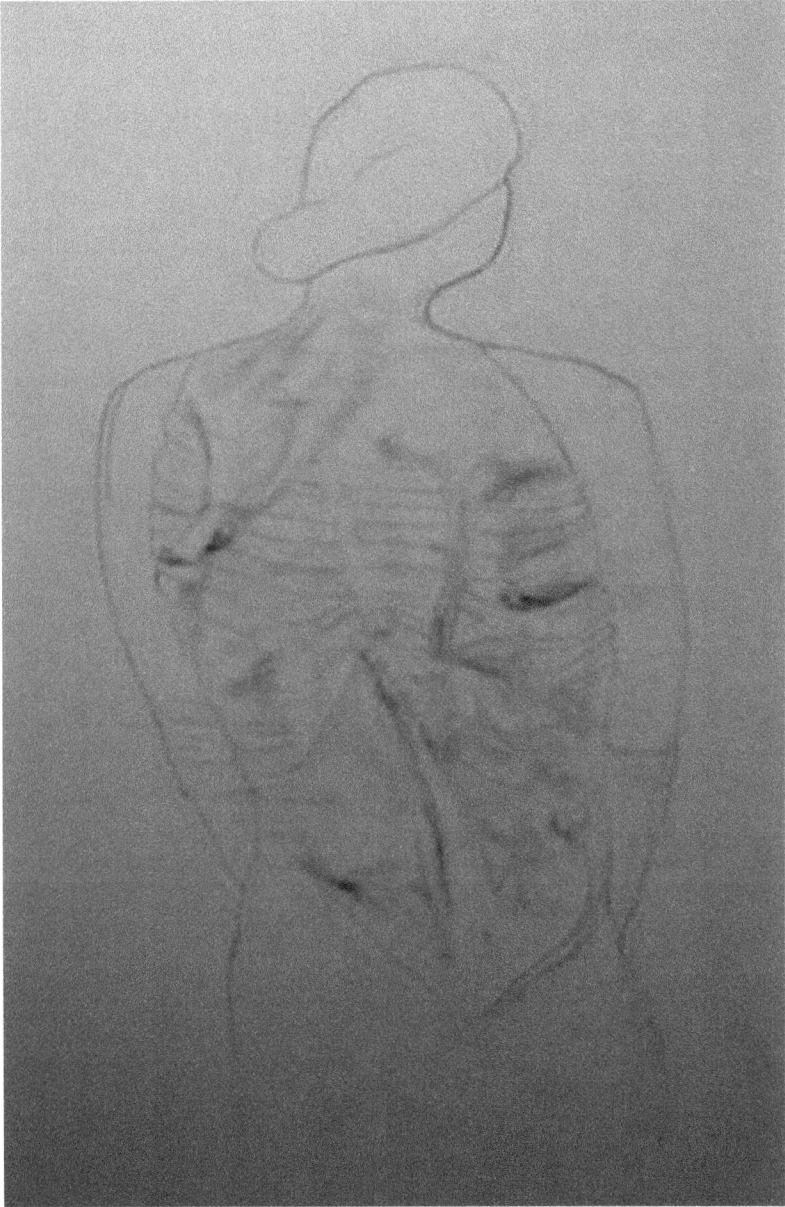

Richards considered the images for a few moments but was unable to offer any convincing insights.

"Well," he ventured, eventually, "It appears to be of two women, or girls, who look the same. Sisters? Twins, maybe?"

"It is certainly amazing how the artist has managed to draw

one as an exact replica of the other, with such a casual free hand, isn't it?" said Merisi, admiringly. "But no, it is not a picture of twins. Merisi described it as a mirror image – a mirror image of oneself. And if you look more closely Dottore, you will see that there are subtle differences in the figures as they do not reflect each other. Their legs lying horizontally do not reflect in the same position. Do you see that?"

"Yes, I can see that. The woman on the left is presenting a different position."

"Although, it is unmistakably the same person, this tells us that the person that you perceive is not a true reflection of that person, even when it is yourself that you perceive! You must be aware of denial and find trust in what you see. You must be aware of who your friends and your enemies really are, and look deeper, so that you can find true identity. People may not be as they first appear!"

"He also said that he had drawn them this way to encourage me to look deeply into my own personality. He felt that it was important for everyone to understand their own self, and to know the difference between the person they are deep down, and the persona that they present to others. You must look in the mirror, Dottore – and I mean really look in the mirror."

Richards followed Merisi as he wandered to the end of the room, and both men had a better view of the final two brightly coloured images.

"These are the drawings that are most important for you, Dottore. This one," said Merisi, gesturing towards a rather abstract sketch that appeared to depict a ghostly apparition of a man hovering above a more solidly constructed skeleton, "represents the art of shape shifting. And of course, you have had some first-hand experience of how effective it can be."

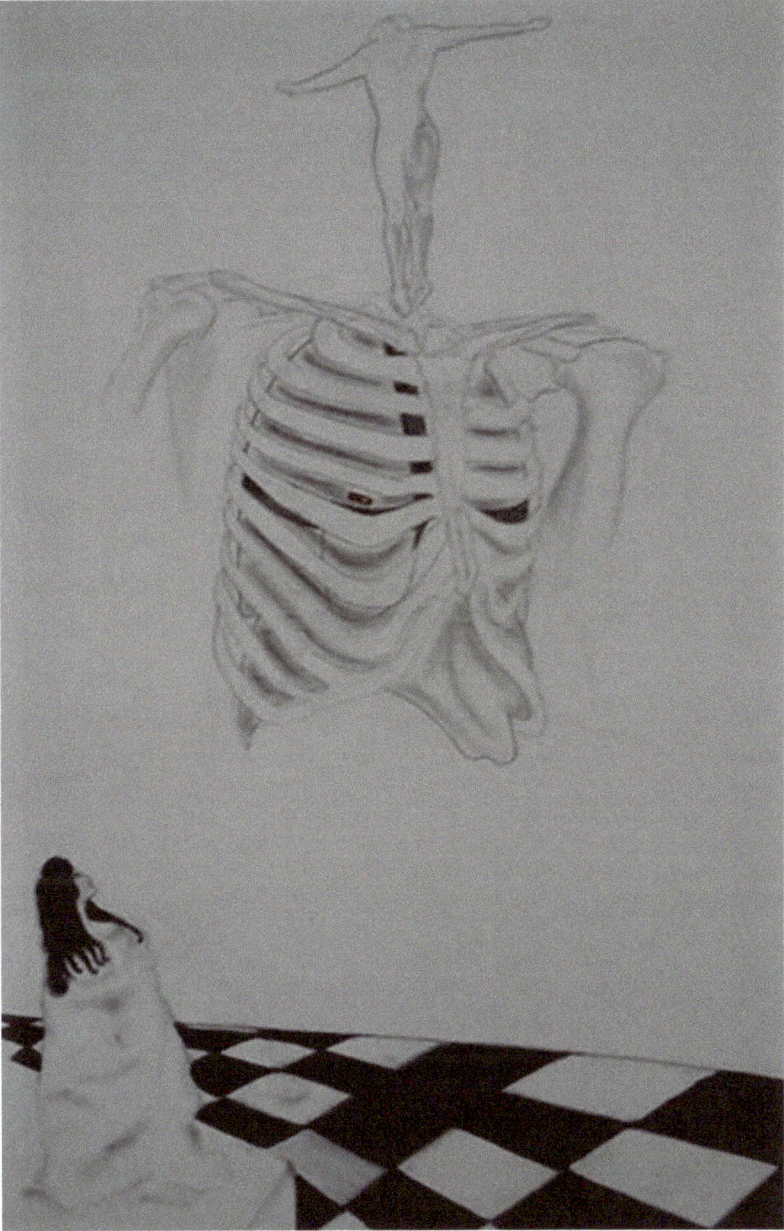

Richards winced again, still embarrassed at having been so easily duped.

"What Caravaggio is telling us in this drawing is that we need to be aware that future technological advancements might be deceptive and dangerous to mankind. We now see this in the form of shape shifting androids coming from the future and taking human form in the present. For centuries power has been achieved through the wearing of armour even from Roman times. In my adapted drawing I have used his prediction that armour will progress to being a device of deception – and shape shifting is the ultimate expression of this with cybernetic organisms using such armour to disguise themselves with a human anatomy."

Richards stood back in amazement at what he was hearing and seeing.

"But this," continued Merisi, "is the image that I want you to study - and I want you to remember it."

The two men stood in the neon glow of a drawing that contained an image of a man who appeared to be soaring above a chaotic world below.

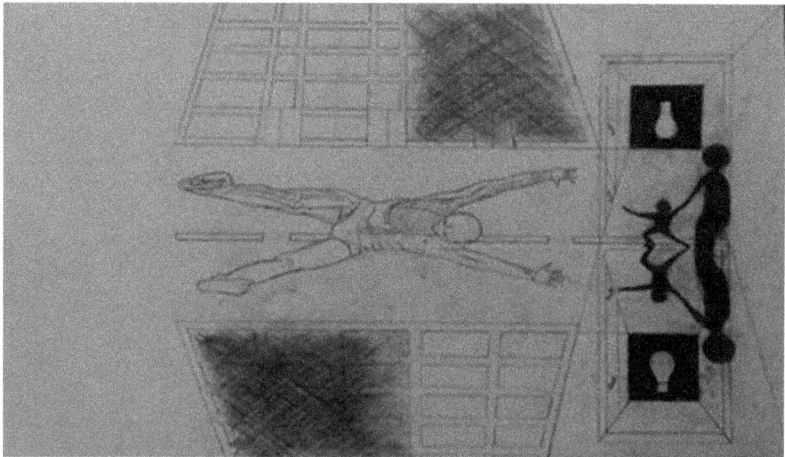

"Caravaggio referred to this one very simply, as 'Volare' – 'Fly!' He wasn't referring to the literal act of taking to the air, but of somehow transcending basic human consciousness in order to have a better view of the world than anyone else. I believe that Caravaggio was able to do this with his brilliant insights into the nature of life and human existence – and I also believe that he was able to paint better than anyone else because he saw things much more clearly. But you, Dottore, will have the opportunity to do something that nobody in history has been privileged to do. You will have a glimpse of the future. In effect, you will be able to see further than any man has been able to see before."

Richards looked dumbfounded at thought of his journey that that lay ahead.

"The great man was also suggesting that we should not blame ourselves for not being able to change an event that has happened in our life, and that we should free ourselves from guilt. This will help us to find self-belief and our true identity, so that we can fulfil our true potential."

Richards could relate to this, as it struck a chord instantly with his own past. He had a vision of himself holding his mother's hand in the picture. He relived the death of his mother and remembered her not being able to hear him shouting a warning to her. He realised that it was not the fact that his voice was not loud enough or convincing enough - it was just bad timing.

"Dottore!" barked Merisi, snapping Richards from his trance. "You see, the true expression of your leadership lies in the development of your ideas. That is your true calling, and that is what the world will recognise as being your true contribution. When making speeches, you must not concern yourself with being loud and convincing to the audience. You must not fear being heard. You must never worry about not being convincing enough for your ideas to be believed. Self-belief is more powerful than what others think of you."

"Caravaggio also explained to me that once we have self-belief and are cared for by a soul mate who believes in us, we can gain self-confidence and a belief that our ideas will be believed by others. Once this has been achieved, we can develop our ideas and our relationships."

Merisi was interrupted by a sudden screech of brakes followed by a loud bang and the tinkling of shattered glass on tarmac – the unmistakable sound of a vehicle colliding with something at speed. Both men turned to face the double doors that they had used to enter the nightclub, for the noise appeared to come from that side of the building. In an instant, the drawings that had filled the room with a multi-coloured light were extinguished, and the pale green 'Uscita' sign over the door became, once again, the only source of illumination.

As Richards' eyes became accustomed to the relative darkness, he saw Merisi had made his way to the door. Merisi's right eye and cheek were then lit by the pale light of the moon as he cracked the door open by an inch or two. Richards remained where he was, standing still and silent, straining his ears for any sound that might hint at what was going on in the world outside.

There were distant shouts – raised voices, which Richards could not understand, although there was no doubt that there was animation and fear in those voices. Then a scream pierced the air. A high-pitched scream of horror, which instantly reminded Richards of Grazia – but his attention immediately returned to Merisi, who closed the door and hurried towards him.

"I have seen and heard enough, Dottore," he said.

"Do you think that Shui Feng is out there?"

"I do not know. But it would not surprise me in the slightest if he were behind whatever is going on – I am not going to take any chances." He took Richards by the arm and led him out of the room and through a series of dark narrow corridors, all of which appeared to be painted in a dark purple colour. As the pair hurried through the night club towards the front entrance, Merisi spoke in a hushed voice: "We need to go. We must get to the ruins of Pompeii and find the portal – and hope that we are not followed. In fact, I am hoping that our friends from GIATCOM have not somehow discovered my intentions.

"There is one thing that I have yet to tell you about Pompeii, Dottore. Something there will be of great interest to you."

"In Pompeii? I am intrigued!"

"If we can make it to the site of the portal undetected, I will show you evidence of just how advanced the Romans were with the generation of energy from wind power. And using turbines that were located in the midst of their cities, driving their mills and furnaces."

"Wow. That's unbelievable...to think that they were-"

"Unbelievable? I think my friend, you will soon be reconsidering just what is, and is not unbelievable – and it will be both the past and future that will amaze you! But, first, you need to get away from this place."

Richards turned to look at Merisi. There was a look of fear and amazement on his face.

"So – you are not coming with me?" he asked.

"No, Dottore. I have a score to settle - so will watch over you while you get away from this place. From here, you need to find your own path, and in the process, free your mind of discontent. And remember - I have given you those yellow drugs for a reason, and I have used my reproduction of Caravaggio's sketches to help you to organise your thoughts, and then fulfil your potential. Good luck, Dottore!"

CHAPTER 9

CLOSE SHAVE

A s Dr Ben Richards hurried through a darkened, cobbled alleyway, towards a glimmer of hope dipped in sunlight ahead of him at the top of the hill. Many appeared to be following in his footsteps. It was now over half an hour since Merisi had bidden him a hurried farewell at the door of the nightclub, and he was now feeling more vulnerable than he had ever felt in his life. He was sure he was being followed - or was this just his instinctive paranoia?

His shirt was wet with sweat, and his mouth dry. He gathered his thoughts and made his way up the hill. He heard hurried footsteps scampering up behind him and the urgent barking of a large dog. The footsteps and the barking gradually became louder – then a small boy with the dog by his side raced past him.

Richards stopped and pressed his back against a brick wall enclosing one side of an alleyway. Breathing deeply, he took out one of the pills Merisi had given him and put it in his mouth, trying to stand very still while his body shook off the fright. When his breathing had steadied, he continued his brisk march.

At the end of the alleyway, he entered a small square. He encountered a few passers-by who were either returning late after a night out or setting off early at the start of their day. He examined each person as they passed him with a skilful eye now that the drug was starting to take effect. But, even with the enhanced awareness that the drug was giving him, he was unable to see behind. His mind kept telling him: *"Don't look back!"* Was there someone following him?

He stopped at a faintly marked pedestrian crossing to await the passing of a municipal refuse truck on its early morning rounds. Feeling a tap on his shoulder, he froze, and half turned causing the first beams of the early morning sun penetrating... a gap between the buildings bordering the square, to laser into his eyes. He became aware of the silhouette of a rather short figure on his left. As he leant back and held up his hand to visor his eyes, almost not wanting to know who it was that had approached him, he was relieved to be greeted by a familiar, smiling face.

"I thought it was you hurrying along in the shadows. What are you doing at this time in the morning?" Father Luigi was wearing a light grey sports jacket over his cassock. His jacket was open, revealing all 33 buttons stitched in a neat row on the front of his vestment. "I am heading for the cathedral to begin the morning mass," he said. "How about you?"

Richards let out a heavy sigh of relief.

"Father! How wonderful to see you! I was just... coming back from the City centre."

"Oh, I see," said Father Luigi, having no intention to pry into where Ben had been, and where he was heading off to – since, having been a priest for almost his entire adult life, discretion came totally naturally to him. "So, have you stopped answering text messages, Ben?" he asked.

Richards was suddenly reminded that he had been without his smartphone since he and Grazia had been abducted outside Caravaggio's restaurant. He was immediately struck by the fact that, with everything that had gone on, he had not missed having his phone.

"Oh, you tried contacting me," he replied in a casual tone, hoping his old friend was not going to make him go into details of why he had not responded.

"Yes, Ben. I have been doing some thinking about the SATOR SQUARE card you showed me when we last met."

As they walked past a caffè, that was open for the early morning breakfast trade, a strong scent of Italian dark coffee filled the air from its open windows. It almost took Ben's mind off his troubles. Almost. But it presented the perfect opportunity for him to take a break from his trek and anxiously looking around him for potential threats. He stopped suddenly at the entrance to the caffè and took Father Luigi gently by the arm.

"I'll tell you what – why not talk about it over a nice coffee?" he suggested. The two men walked into the welcoming warmth of the Caffè Taranto. *"Due caffè Nero,"* he requested, in his basic Italian. The barista squinted at seeming him a little confused, but ultimately understood him. At least he hoped she did. As the coffee making machine roared into life, Richards emptied the contents of his trouser pockets, including, among other things, about €20 in coins and assorted crumpled notes, onto the counter.

"Please," insisted Father Luigi, "let me pay for these-"

"No, no," insisted Richards, turning and gently preventing his old mentor from producing his wallet from his jacket pocket. It was a well-meaning gesture from the young engineer, but one that was to prove costly as it caused him to fail to notice the cylindrical tube of tablets Merisi had given him rolling off the counter. It settled amongst a multi-coloured bowl of sweets.

As he waited patiently for the barista to complete the process of coffee preparation, he wondered just how much he should tell Father Luigi about his strange, and scarcely believable adventures of the past few days. As the young lady placed two coffees on the counter, Richards turned to Father Luigi and said: "Could you find a table for us? I just need to visit the loo."

After relieving himself in the cool, dimly lit, but surprisingly spacious, facility, Richards washed his hands, then tried to revive his tired mind by continually splashing cold water onto his face. As he did so with his eyes tightly shut, the gentle sound of the water running from the tap and tinkling into the basin was suddenly drowned by the racket made by the automatic flushing system of one of the urinals, which suddenly sprang into life.

The sudden increase in the volume of noise, and Richards' prone position as he bent over the sink, provided the perfect opportunity for the dark figure that now emerging from the darkness of the nearest of the three compartments in the room.

The figure raised a hand, which held a small hypodermic syringe high above its head as though it was about to plunge a dagger into Richards' neck. The crude attempt was thwarted, a man dressed in white, quickly and silently entered the room and seized the would-be assailant by the wrist holding the syringe. It was Merisi. With a swift kick to the back of the attacker's left knee, he was soon able to wrestle the faceless creature to the tiled floor of the toilet, and with a strong left arm across the throat of the TX-series android, soon rendered the contraption lifeless. Merisi had been distracted from his errand when he noticed what looked suspiciously like a GIATCOM android follow Richards away from the nightclub. During his long career as an UNA agent, he had become very familiar with this class of robot, and was able to recognise the very direct, rather stilted manner they had when stalking their prey. He had also had a great deal of experience with intercepting and disabling GIATCOM's weapon of choice, so he made short work of his adversary.

Once the robot was motionless – and with Richards continuing to splash water on his face, oblivious of what was going on behind him – Merisi dragged it into the cubicle from which it had emerged. He examined the syringe he managed to prise from the android's grip with some difficulty. The liquid it contained was aquamarine in colour. Merisi recognised it as a substance commonly used to render someone unconscious for several hours – and the deep colour of the liquid suggested a strong concentration. Merisi had already received intelligence from UNA's HQ it was GIATCOM's intention to apprehend Richards, rather than kill him. But why? What was their plan for a man whom he, himself, was trying to protect in order to ensure the world would have a future without GIATCOM, and who had, in a now-rewritten version of history, been assassinated?

Merisi had been sent on his mission with the express remit of protecting Ben Richards from a death that had already occurred – yet, there was now every indication the very same organisation, contrary to thwarting him in his task, appeared to have a very similar objective.

What could GIATCOM want with Richards? An engineer and budding entrepreneur who was totally committed to creating a future based not on nuclear power, but on renewable energy sources? What was GIATCOM's strategy?

These questions troubled Merisi greatly, but he was distracted from his thinking when the sound of running water suddenly ceased. Out of sight, in the cubicle with the door half closed, Merisi listened as Richards walked a few paces to the hand towel hanging on a hook beside the entrance to the toilet. Moments later, Merisi heard the high-pitched whine of the door's hinges as Richards left the room. It was followed by a creak of a much lower tone, as the fire door closing mechanism forced the heavy door to swing to a close behind him.

* * * * *

Father Luigi smiled as Richards approached him. It was an almost child-like smile of delight – and the reason for his exaggeratedly good humour was the presence of a small glass half-full of a clear, yellow liquid placed next to his coffee.

"Look, Ben," he exclaimed. "I have been given a complimentary *digestivo*. You could have had one as well – but I know you are not too partial to that sort of thing. I never could interest you in the subtleties of the *aperativi* and *digestivi* that we have here."

"No, Father, I'll stick to my beer and my wine."

"Apparently, they're trying a new brand, or something. It's good. It tastes a little like *Ricard* – but less sweet."

The grey-haired clergyman smiled again after he had taken another fond sip of the anise and liquorice-based concoction, but his expression changed to something grave once his young friend and onetime foster son sat down in the iron and wicker chair opposite him.

"Ben," he began, "the reason that I have been trying to contact you is that I have been thinking about that SATOR SQUARE you showed me..."

"Yes, you were saying – and you sent me a text about it?"

"Yes, because I'm a little concerned for you."

"Concerned? Why?"

"Well, you seem very interested in the whole subject area – I am sure that it is something that has piqued your curiosity –

which is no surprise, you were always fascinated with puzzles and mysteries – but a SATOR SQUARE is not something that should be taken lightly." He paused, and looked closely at Richards, to ensure that he had his full attention. "This is not merely some ancient game of Sudoku, designed to amuse or pass the time."

Richards could tell that his friend was genuinely troubled. "What do you mean, exactly?" he asked.

"You will remember I told you the Squares are known to have been used by the earliest Christians?" Richards nodded. "Well, they were later often used by witches and soothsayers. Ben – women were sometimes burnt at the stake for engaging with these things."

Father Luigi looked into the young man's eyes, and Richards again noticed the air of concern, but now it was tinged with that expression of genuine affection that he had known so well in his youth, when the old pastor was always there as a source of support and encouragement. The subject of witches and witchcraft usually drew some flippant remark from Richards, whose keen scientific brain generally dismissed such things as being totally absurd, but there was something in Father Luigi's manner that suggested that he should take his advice seriously.

"And," added Luigi, "I once lost a very dear friend who had a particular fascination for these Squares. Whether she passed through, or fell into a time portal, I do not know. All I can tell you is that she went to examine a particularly interesting SATOR SQUARE one day – like you, she was full of scientific curiosity and excitement – and I never saw her again. Not a single day has passed without me praying for her return."

Richards was at a loss to know what to say to the priest. He was, of course, particularly struck by Father Luigi's reference to this 'very dear friend' as 'her.' Furthermore, Luigi had never before confided such personal information to him.

"Yes, Ben," continued Father Luigi, after a short pause. "You are not the only one to have suffered loss in your life."

Richards decided it would be inappropriate to delve too deeply into the precise nature of the relationship between his friend and 'her,' so he refocused the conversation on the issue of SATOR SQUARES.

"So, what did they use these things for?" he asked.

"Most often as a warning to people who had something to protect," replied Father Luigi.

"You mean their fortune, or their wealth..."

"Maybe. But not just material wealth. Very often, it was the protection of knowledge that was most important for people. It might be some valuable piece of military intelligence, or a commercial or industrial secret enabling a city or religious order to maintain dominance over rivals. I am sure the ancient world was, and still is, littered with stashes of manuscripts and other sources of knowledge and wisdom. It was particularly commonplace in ancient China, for example. Prior to going to war with a rival warlord, or maybe when defeat appeared to be inevitable, the practice was to dump all evidence of the knowledge that had helped each society prosper, in the deepest lake or at the bottom of the deepest mine."

"As I explained to you in the cathedral, the meaning of the SATOR SQUARE is not known for certain. Scholars have suggested a number of translations from Latin, but the general message from the five words making up the Square appears to be 'the sower controls the turning wheel.' In one sense, it is a prophecy suggesting the originator of a resource, or an initiative, or an idea is the one who controls outcomes, good or bad – but, at the same time, it is a warning there will be others who will try to wrest ownership from the sower. The rationale behind stowing away a cache of knowledge and other valuable information is it can be re-used when the cycle of power comes back in favour of a particular society or sect. Even if all the artisans and scholars were murdered, their knowledge and know how can be rekindled. The pursuit of these less lucrative treasures is a very under-rated branch of archaeology."

"Maybe the person who gave you the card bearing the SATOR SQUARE is trying to warn you to protect what is yours. Who gave you the card?"

Although he implicitly trusted his old friend, Richards instinctively knew it would be best not to make any mention of Merisi, or of the wider and more consequential issues of which he was aware.

"Oh, it was just some old bloke at the conference," he replied.

Richards could tell from Father Luigi's expression that he was far from convinced by his attempt to play down the importance of the stranger giving him the card – but the pastor was too polite to press Ben further. "I don't know what you are getting yourself into, my son," continued Father Luigi, "but be careful. Passing someone a SATOR SQUARE is rarely done for no reason at all. I will certainly remember you in my prayers tonight."

The two men fell silent, with Richards not wishing to elaborate further for fear of inadvertently divulging something about his wider mission Merisi had insisted he should keep to himself, and Father Luigi was satisfied he had performed his duties as a friend by warning Ben to be careful.

It was the elderly priest who eventually broke the silence: "So – where are you heading off to at this late hour?"

"I am actually making my way to the ancient ruins of Pompeii," Richards replied.

"Ah! Hunting for more SATOR SQUARES, no doubt!"

"Yes. Something like that," said Richards, with a slightly nervous smile. Then it occurred to him that Father Luigi's suggestion might have been more than mere guesswork. "Why would you associate Pompeii with SATOR SQUARES?" he asked, attempting to feign idle curiosity.

Luigi smiled briefly before answering the young man's question.

"I have probably been less than honest with you, Ben," he confessed. "I know a great deal more about the location of SATOR SQUARES than I have so far been prepared to admit."

Richards looked at his old friend, as Father Luigi drained the last of the yellowish *digestivo* from his glass. He realised it was pointless to continue trying to be coy about the purpose of his trip to the ancient site – and he was keen to learn as much from his old friend as he could.

"So you know about the portal in Pompeii?" he asked, "and my reason for going there?"

"I know enough," said Father Luigi, nodding sagely. "And the rest I can probably guess. I don't know why you are going there, or on whose behalf – but I do know a great deal of knowledge,

as well as many secrets were buried under tonnes of ash on that
fateful day. And, yes, one of the most important secrets 'lost' was
the existence of a time portal."

"The SATOR SQUARE is the perfect example of imaginatively
using an archaeological and ancient relic of worship that can
be used as a time travel portal, due to its mystical, puzzle and
palindrome qualities. It implies that time can be reversed or
moved forward, as symbolised by the words being able to be read
backwards and forward."

"The word TENET is a palindrome that can be read the same
backwards and forwards and could signify that time could be
changed in the past or the future which impacts the present due
to a change in the past or future. TENET can be read in reverse
in the form of a spiritual cross formation, which adds to its
mystical qualities and suggests that time travel is spiritual. My
own interpretation of the words is - He who sows the seeds, has
the key to the wheel of life. Making me realise the words were
referring to - Creator / Saviour with its religious origin."

"I was fascinated by how all the words spelt together to form
a square, whilst being able to be read the same forward and
backward. Specifically, the middle word TENET which not only
read the same forward and backward but the same up and down,
which represented not only the spiritual symbol of the cross but
also symbolised time in the form of a clock face. Being able to
be read the same at positions 12, 3, 6 and 9 on a clock face. This
gave me the idea of time travel being linked to spiritualism and
the church hiding the secret for generations."

"As religion is mystical, it occurred to me that this magical
square represents religion as being mystical, given the fact the
SATOR SQUARE is mystical itself in being a puzzle."

"So the word TENET was a code word to symbolise time
travel and let certain Christian or spiritual people of that time
know that there was a time travel portal hidden by the church?"
asked Richards.

"Yes."

"I believe it was his in-depth knowledge of Egyptology that
made me think the SATOR SQUARE had an Egyptian influence,
although the language of the square is written in Latin. Which

indicates it coming from the Western part of the Empire at the time of its creation. I believe the occurrence of acrostics and crosswords within Egyptian literature made an Egyptian influence on the origin of the SATOR SQUARE even more likely. We know that the first sign of the square being found was in Pompeii AD 79."

"When St. Paul landed at Puteoli, AD 60, he found Christians there to welcome him and found Egyptian cults had been established. Pompeii too had its part on the River Sarno, and traffic with Egypt was lively. It is known of course, that Pompeii also had a flourishing cult of Isis. It was influenced by Alexandria, a city which fostered not only Egyptian cults but also at an early date, Christianity."

"Of course, Pompeii as a source of ancient knowledge is different from the others, in as much as the eruption of the volcano was unexpected. That means the manuscripts and artefacts were not carefully 'hidden', so, in principle, they should be easier to find – but you would still need to know where to look for them. Of course, many objects of interest from Pompeii have been excavated, restored and studied, and are now on display to tourists, but not all of them. What particularly attracts you to Pompeii, Ben? I sincerely hope that it is not the old dream of time travel."

Richards again considered how much of his mission he should divulge to the priest.

"I'm just going there to see what I can find there," he lied, not with a great deal of conviction. "It will be interesting to see if I can find any SATOR SQUARES, of course."

Richards detected a frown on the face of Father Luigi. It was a grave expression, which he initially took to be one of disapproval, but, as the priest's brow became furrowed and the skin around his eyes started to redden, Richards quickly realised that his friend was suddenly in considerable pain.

"Father?" he asked anxiously. Father Luigi opened his mouth, as if to speak, but was unable to utter a word. A choking noise issued from his mouth as he clasped a hand to his throat, tearing off his collar with the other hand.

"Help!" yelled Richards, desperately glancing around the

deserted caffé. "Aiuto! Aiuto!" But it seemed that nobody was within ear-shot.

Within a second or two, Father Luigi's head had taken on a beetroot red colour. Even the choking noise from the old man's throat was now barely audible, as he fell forward into Richards' waiting arms. *"Aiuto!"* called Richards, randomly into the air, and this time with less conviction, as it was now obvious that nobody was going to come to his friend's aid.

Richards felt the warmth of the blood that gurgled quietly from Luigi's mouth, seeping into the fabric of his shirt and his trousers. Then his friend's body stiffened slightly. Then all was completely motionless.

Richards' senses were suddenly filled with grief and horror. Yet, to his surprise – he was unaware whether his mind was being affected by the drug that Merisi had given him – he retained the ability to think. Immediately, he reasoned that Father Luigi must have been poisoned. What's more, this was clearly an assassination. The only perpetrator could be a GIATCOM agent who had seen Father Luigi with him.

Richards glanced accusingly at the small glass that had contained the yellow *digestivo* – but there was no time to ponder the precise cause of his friend's death. If an agent had been able to murder his friend, then it was obvious they could not be far away. Suddenly, the paranoid sense of being followed that had haunted him during the evening had returned. He resolved that he must tear himself away from the body of his dear friend and immediately continue on his journey.

In an apparently callous act – one which would have appeared to a casual observer to have borne all the callousness of a murderer – Richards allowed Father Luigi's lifeless form to fall onto the tiled floor of the caffé. Just seconds after his close friend and former mentor had drawn his final breath, Richards was once again walking purposefully through the mild early morning. He walked with his head bowed and with the collar of his jacket turned up. As he left the caffé, he had grabbed a handful of paper serviettes, and had used them to mop up as much of the blood on his clothing as he possibly could and dumped the handful of paper into a bin as soon as it had become too soaked to be absorbent.

Richards hurried along with a single-minded gait; his eyes fixed on the pavement immediately in front of him. There would be a time to mourn the loss of Father Luigi – and, unbeknown to him, there were to be many empty days and weeks during which he frequently cast his mind back to the time that he had spent under the priest's wing as a young boy – but, for now, Richards' mind was bent solely on survival in the present. He never looked behind him, or to either side, as he made his way through the deserted streets. He resolved to not stop again until he had reached his destination – even if *The Pope* himself called out to him and asked him the time.

But strangely, in spite of having a feeling of impending and immediate danger, he found his mind wandering from the current imperative to 'get away'. With only the monotonous sound of his own footsteps echoing in the early morning silence, he found himself thinking of Grazia. At first, it was a pleasant recollection of her deep brown eyes and her impish smile. Then his thoughts turned to guilt that he had somehow involved her in this perilous adventure. He deeply regretted this young woman's life had been put in danger, as part of something that he still barely understood, but which obviously seemed to be very much focused on him. He felt utterly helpless that Grazia was in the hands of some very dangerous and unpredictable people – whose very existence was a challenge for him to comprehend – and he found it hard to come to terms with the fact they were probably both entirely reliant on Merisi for their safety and for their life.

His racing mind briefly wrestled with the nature of this man, if that is what he was, who called himself 'Merisi.' Who, or what, was this character who he had so quickly learnt to trust? Not for the first time, there was a part of him wondering whether everything had just been a bad dream, or one of his hallucinations, and he would soon suddenly find himself awake again, alone in a hotel room, preparing for his next presentation. But, as he turned right into an unevenly paved side street, the steep slope and the cooler breeze greeted him appeared real enough.

Very soon, his thoughts returned to Grazia. This time, he was reminded of the brief time they had spent together in the apartment in Milan. Specifically, he recalled a conversation they had had over a morning coffee. Grazia was telling him about the brief time that she had spent as a trainee in a *Carabinieri* Special Forces unit; her originally chosen career. However, she abandoned that path when she was assigned to a mission in a mafia-controlled part of Palermo. Grazia opened up to Ben how she became very close to an operational partner who, due to a gunshot wound to the head, had dysfunctional amygdala-the area of the brain that supports decision making in stressful situations. This meant he lacked the ability to sense danger, and Grazia, with her very good judgement in such situations, became an important part of his life.

Grazia confided to Ben that the mission in Palermo had been her last, because her partner lost his life during a shoot-out, he should not have got involved in. It had been part of her job to steer him away from such situations, and she felt very responsible for his death.

Maybe it was the sense of being in danger that lurked in the back of his mind, causing Ben to think of Grazia in this way but was reminded of how she had sensed the threat posed by the motorcyclist who followed them on the journey from Florence. It was not something that would have occurred to him if he had been alone, due to his natural, boyish naivety. He soon began to realise how much he missed her, and how much he wished she were with him right now. In fact, it soon became very clear to him how much he needed her – almost as much as her ill-fated partner had needed her.

As the morning sun began to cast sufficient light to make the skyline of the city's residential buildings just visible on the horizon, Ben had the very clear revelation that there was something in Grazia's personality that made her need to feel needed by another person. That made him long for her even more, as he trudged through the empty streets.

CHAPTER 10

ON TOP OF
YOUR GAME

SEPTEMBER 4, 2017

Tony Clarke made his way to the departure lounge of Roma airport, pulling a suitcase rolling along behind him with one hand and carrying a half-full cup of coffee in the other. He felt low and empty, not knowing what to do about the troubled finances of his football club. But his spirits lifted when he reflected upon the time spent in Italy, recalling his chance meeting with Ben Richards, inspiring him with his innovative ideas. He was taking up almost two seats in the lounge with his backside, which resembled the width of a fireplace, to the discontent of a rather glamourous Italian lady wearing shades sitting opposite him. He could see she was in some sort of discomfort as her hand appeared to be held tight by her accomplice, a peculiar-looking man with curly, ginger hair who whispered in her ear in a Russian accent. Tony could feel the heat radiate around the collar of his shirt as Vulkan's eyes addressed him and exploded to a devilish squint. As Tony turned his eyes towards the coffee table, the woman took off her shades with her other hand to reveal her familiar face. "Hey, don't I know you from somewhere, madam?" She shook her head and looked away, trying to look nonchalant as Vulkan began to squeeze her hand tighter.

Clarke picked up an Italian football magazine called NESN Soccer, which caught both his and her eye with a picture of a stadium roof on the front cover. He opened the magazine to a double page spread on to the table so he could see through his far-sighted varifocal lenses. To his amazement, an article about Dr Ben Richards' keynote speech on one side, and a story of AC Milan selling their best players to avoid bankruptcy on the other.

The article read as follows:

The recent sale of two AC Milan's top defenders to Paris Saint-Germain for £49 million reflects AC Milan and Italian Soccer's new Economic Reality.

We know it's a new day and age when AC Milan can't hold on to its best players. But that is the world in which we now live.

Selling the world's best defender and best striker (arguably on both counts) strips one of soccer's most prestigious clubs of some of its lustre. But it brings Milan closer to compliance with new financial rules changing the face of club soccer in Europe.

Adopted in 2009, the Financial Fair Play (FFP) regulations prohibit clubs from spending more than they earn. FFP rules are being implemented over the next three years, but UEFA watchdogs are using accounts from the start of the 2017-18 season to judge whether clubs are operating with "discipline and rationality."

Clubs are found to be in violation of FFP, UEFA can bar them from entering lucrative competitions like the UEFA Champions League and Europa League.

Former Italian prime minister Silvio Berlusconi has owned Milan for 32 years, and he's used some of his vast personal wealth to help fulfil lofty ambitions (both Milan's and his own).

For a generation, Milan has fielded star-studded squads and paid some of the highest salaries in world soccer. The club has become used to winning under Berlusconi, taking home ten Serie A titles and coming second six times. It also captured six Champions League trophies and finished as runner-up four times.

Nobody seemed to care that the club has been bleeding money for years - the club has lost an estimated £51 million in both 2015-6 and 2016-7 seasons — or that Berlusconi's generosity was helping to keep the club near the pinnacle of European soccer.

But UEFA changed the rules, and even Milan must comply with the new ones.

Forced compliance meant that Milan had to cut down on its hefty payroll, which was estimated at £151 million in 2016-17.

A whole host of AC Milan's team were sold at the start or allowed to leave before the 2016-7 season.

The departures trimmed the payroll by roughly 40 percent, but also caused considerable unease among Milan players and fans. In the last month, current players have publicly pleaded with the club not to sell one of their current best players. Without their top defenders, Milan has no reference points in defense.

Milan's revolution has begun, and it coincides with one going on in Italian soccer. Financially, the Italian game has fallen behind the English, German and Spanish leagues. Game day revenue lags way behind other leagues, as increasing numbers of Italian fans are staying away from crumbling stadiums. Also, a new, more equitable system of distributing television revenue has cost Milan a chunk of its income at a time when big clubs in other countries are receiving more television money than ever.

He was transfixed by this article and how coincidental the link between troubles with stadium finance on a global stage and how a chance meeting with Dr Ben Richards could be fate. Telling him he needs to pursue the opportunity of innovative engineering, to help cost savings on powering the stadium and create additional revenue from selling the power created as a commercial venture. So that Football clubs can hold on to their players rather than sell them off as assets.

"Glasgow Rangers FC was sold for a penny, due to financial ruin!" He said quietly to himself, trying to stop his hands from shaking the magazine.

What is more coincidental is in the magazine Dr Richards'

story of innovative engineering precedes the story of Football Clubs having to sell their players to avoid financial ruin? A sign, Tony thought, as being 'Prevention before Cure'. By using Richards' innovative engineering, one could prevent selling prized assets to cure financial woes and make revenue gains from saving on stadium electricity costs and selling the export electricity generated from the stadium roof to nearby businesses and residential housing. At the bottom of Dr Richards' story was a picture of the journalist, Grazia Rossini, posted 2nd September 2018 14.12pm. He immediately looked up and gazed deep into her eyes. "Kind of fortuitous, don't you think!"

"Fortuitous? ... Coincidence more like!"

Vulkan squeezed her hand once more and rose to his feet, pulling Grazia away from her seat.

"Ow! Let me go, you brute! *OustiaPouta!*"

As he dragged her across the airport lounge, she turned her head to the left, and pointed her finger to Tony as she looked back, then made a telephone shape with thumb and little finger holding it to her left ear. Tony understood what she was trying to say, as she dropped a little white card from her left hand on the floor behind her.

CHAPTER 11

ANCIENT WISDOM

Ben Richards fumbled in his wallet for an additional €20 note, as the taxi driver waited patiently in the in the amber glow of a nearby streetlight with his forearm resting on the top of his steering wheel. The fare from Naples Airport was rather more than Richards expected, but he was happy to have arrived at his destination.

"Mille grazie," he said, handling over a wad of notes to the driver. With a nod, he had now perfected. Both men understood Richards was not expecting to receive change.

The driver stuffed the notes into the breast pocket of his burgundy shirt. "I wish you luck with whatever it is that you want to do," he said, demonstrating his English was much better than Richards' Italian, "but I am telling you – it is all closed now. All locked up."

"OK. But there is something that I want to look at," replied Richards. "Something I want to see."

The taxi driver said no more. He merely raised a broad palm, put the car into gear and swung away, heading off in the direction of Naples.

Richards took a deep breath and surveyed the scene. He already knew it was just past 2am, even as he glanced at his watch shortly before the taxi arrived. He felt tired. He had only managed a few minutes sleep on the plane – it had been a slightly bumpy flight, on the 100-seat Embraer E190 – but the adrenalin pumping through his veins was keeping him alert.

All was quiet once the tinny sound of the taxi's engine disappeared into the distance, all was quiet. He could see the gentle curves of Mount Vesuvius, a silhouette on the horizon made visible by a combination of the clear, starry sky and the light pollution from the nearby city. He could also make out a few truncated stone columns in the middle distance, which contrasted with the rolling outline of Vesuvius. They were immediately recognisable as some of the many roofless ruins of buildings from one of the world's best known natural disasters, from the year 79AD. Now, it was one of Italy's most popular tourist attractions – so the immediate task for Richards was to work out how to get through the inevitable layers of security. He had no plan. No clue of how he was going to approach his task – but something spurred him on to walk forward and approach the main entrance to Pompeii.

Reassuringly, there were direction signs everywhere. Arrows with icons, pointing the way to toilets and restaurants and gift shops. Information was even available in English: 'UNESCO World Heritage Site;' 'Guided Tours;' 'Archaeological Zone.' But nothing that was of any help to him with his quest. When he drew near to the main entrance with its turnstiles, he was astonished to discover that the large wooden gate in the centre of the tarmacked entrance had been left ajar. What is more, he was surprised at the apparent lack of a security presence at the facility.

Very tentatively, he peeped through the opening, and saw a glass kiosk labelled 'Security / Sicurezza.' By the time he had walked within a few yards of the kiosk, he could tell that the kiosk was empty. Suddenly, the silence appeared eerie –the mood was shattered when a man's confident and firm voice, rang out, just a few feet behind him: "If you are looking for the security guards, Dottore, I have found them bound and dumped behind some refuse bins."

The shock of the man's sudden appearance and familiar voice

made Richards instinctively clasp his heart with his left hand and looked up to the heavens as if in deep shock.

"Merisi!" he exclaimed. "You gave me the shock of my life! Where did you come from?"

"It's bad news," said Merisi, ignoring the question. "Clearly, we are not the first to arrive!"

Richards quickly regained his composure and started to 'put two and two together.'

"Oh! I see! So, the security guards-"

"Yes. Eliminated. Quickly and efficiently. It must have been the work of the GIATCOM team."

"But – the security people. You say 'eliminated'?"

"Yes. Rendered unconscious – and they will be for some time."

"The work of your friend, Shui Feng?"

"Maybe. I don't know. But I didn't want to risk you going into this place alone, so I have been waiting for you."

"Well, I'm very pleased to see you!" exclaimed Richards with a relieved smile.

"So, how do you feel? Are you ready to press on?"

"Yes – as ready as I can be, I suppose."

"Good – because we don't have much time."

"So – what's the plan?" asked Richards, after a slight pause. "There is something here I need to see. Do you know where I need to look?"

"I don't know exactly, but I have a rough idea. Come over here," he added, beckoning Richards over to a large plan of the attraction positioned just inside the long row of turnstiles. The site map was extremely opportune, especially since it was complete with a *'Tu sei qui /* You are here' sticker. "This is where we need to go," said Merisi, tapping a sun-tanned index finger on a scale representation of a rectangular villa.

"'House of the Vettii,'" said Richards, reading the label for the building on the plan.

"That's right. That's where we will find the ancient drawings, I was telling you about – according to the latest information I have. It's not far. About ten minutes' walk – but we will have to be careful and keep our eyes open. Evidently, GIATCOM is here, but I do not know how many operatives they will have sent. You remember those little yellow tablets that I gave you?"

"Yes, I have them here," replied Richards, reaching into his jacket pocket. His face dropped, as he realised that the tube of pills that Merisi had given him was no longer where he had put it. "Hang on a minute..." he said, as he frantically checked all of his other pockets. But it was no use. He clearly no longer had the tube. "No, they're not here. They must have fallen out of my pocket in the taxi, or in the plane, or-"

"Never mind," interrupted Merisi, not wishing to dwell in the well-lit entrance space for longer than was necessary. "It would have helped a great deal if you had been able to take one of those pills right now, but we will have to go as we are." He sighed and reflected for a second or two. Then, he said: "I suppose the worst-case scenario for those drugs is that some innocent drug abuser is going to go on the trip of his life! But we cannot worry about that now."

"Stay close to me – and do not make a sound. Come on – it's this way..."

Once again, Ben Richards found himself focusing on the left shoulder of Merisi's white jacket, as they made their way through the shadows of the ruined city, just as he had done when the two men left the hospital, except that this time, the pace was slower and more circumspect. Both had to tread carefully as they negotiated the mosaic of large stones that made up the narrow streets of Pompeii. Richards was struck by how similar the streetscape was to what he could remember of the alleyways, side streets, and courtyards of Milan – except the partially destroyed walls of the ancient city invariably came to a jagged end clearly visible against the night sky just a few feet above their head. The young engineer was also acutely aware of the irony of the apocalyptic urban landscape around them, considering the images of the future destruction of the world Merisi had painted for him.

Suddenly, Merisi stopped, extending a forearm to halt his companion.

"The House of the Vettii," he whispered. "It's over there." Merisi was pointing to a townhouse, some 30 yards ahead on the opposite side of the narrow street. It was a building that appeared, from the men's vantage point, to have two fairly complete stories – but, like every other building in the city, was roofless. Even in the poor light of the early hours, Richards could see that the walls were fairly featureless, and that the house was constructed of small, almost pebble-like bricks, about half of which were still covered in a grey render. As an engineer, he marvelled at the straightness of the walls and at the sharpness of the corner nearest to them, and was accentuated by the cold light of the moon. More importantly, he noticed there was no apparent entrance to the building. The stark walls showed nothing more than a row of small windows on what must have been the first floor.

"We need to get in there, somehow," continued Merisi in an even quieter whisper. "In fact, we need to go down into the cellar. That's where we should find the time portal."

"Should?" repeated Richards, rather concerned that he might be risking his life venturing into a building which only 'should' be the right one. Merisi immediately detected the air of uncertainty in Ben's voice. "I can only go by the information I have been given – but I will go and have a look, to make sure. You stay here. I can make myself less easy to detect than you can. And don't move!"

Richards watched as Merisi made his way along the shadowed side of the street, the white making him just visible. Then, he felt his eyes were deceiving him; Merisi's suit appeared to actually darken as he crossed the stone cobbles into the light. Richards blinked, and then opened his eyes wide as what now appeared to be a pale, strangely shimmering figure reached the end of the House of the Vettii, then disappeared around the corner.

Richards sank down to his haunches with his back pressed against a hard, cold wall, and wondered how long it would be before Merisi returned – or indeed, how long he should linger in the shadows before advancing from his hiding place to investigate what was happening.

He was alone for no more than ten minutes, but his mind still had time to consider all the events that happened to him over the past few days – and here he was, apparently just hours after his 'death', squatting down in the city of Pompeii, in the middle of the night, waiting around for some seemingly supernatural man who just said "Wait here." He smiled to himself, considering how astonishing his own reality had become – but, then, he was reminded of his sadness at the uncertain fate of Grazia, and his feelings of guilt and helplessness at not being able to save her from being taken away. He also felt an acute sense of personal loss, a lonely figure hunched in the darkness.

Richards was gazing up at the sky watching a cloud obscure the lower half of the crescent moon, when he felt Merisi's hand on his shoulder.

"It is definitely the right place, Dottore," he said, in a hoarse whisper and slightly out of breath, as he knelt down beside Richards. "GIATCOM agents are definitely in the building – so there is no longer any doubt."

"Is Shui Feng there? Did you see Shui Feng?" asked Richards anxiously.

"No, I saw nobody from GIATCOM – but I heard voices. I just hope the voices were not coming from the cellar. We have to hope against hope they do not find the drawings before we do, and more importantly they do not know the exact location of the portal."

Richards allowed his head to drop, as a difficult task appeared to have become considerably harder.

"But I have some very good news for you," continued Merisi, inclining his bearded face very close to Richards' left ear. "I have seen Miss Rossini!"

"Grazia!" cried Richards, suddenly jerking his head up.

"Shhh! Quiet! GIATCOM's people and androids might be anywhere!" demanded Merisi, casting anxious glances up and down the street.

"You saw Grazia? Where? How was she? Is she okay?"

"Yes, the signorina appears to be fine. Her hands were bound, but otherwise, she seems to be okay. I saw no sign of her captors, so I am not yet sure of how many we are dealing with."

Suddenly, Richards' only thought was to see Grazia again, and rescue her, if he could. "So, what are we going to do?" he asked anxiously. Merisi paused for a few seconds, weighing up all the options and all the risks. "Well, we need to make a move," he said, at last. "What worries me is I am sure that Shui Feng will be in there somewhere. The callousness with which those security guards were dispatched, with no attempt made to make it look like a robbery, or an accident, is a hallmark of his – but we need to go in."

"Okay, so what do you want me to do?" he asked, barely able to keep his voice down to a whisper.

"Just stay here until I tell you to come. I will go ahead and check that all is still clear." With that, Merisi hurried off to the corner of the building, crouching slightly as he ran, and disappeared once again.

This time, Richards kept his eyes on the corner of the street. Soon, the slightly hazy outline of Merisi reappeared. Richards could just about make out that his accomplice was frantically beckoning to him, and so he set off immediately. As he turned the corner, Merisi ushered him around the slumped figure of what looked like a large, muscular man – and then there was another, in a similarly crumpled heap, some twenty yards ahead close to what appeared to be the entrance to the house. Clearly, Merisi had been busy eliminating GIATCOM androids. "Snatch team," he explained as Richards stared at the evidence of his work.

The House of the Vettii had been owned by a wealthy Roman family, and although the front entrance was relatively unspectacular, it immediately led to a magnificent atrium. Although its partial roof was destroyed in the ancient eruption of the volcano, at least part of every column bordering theatrium on all four sides remained, providing the two men with the cover they needed to penetrate the ancient ruin.

They scampered from one column to another, pausing to look and listen for GIATCOM personnel at each one. By the time they had reached the fifth column, Richards could see Grazia sitting on a stone plinth, her hands tied behind her back. Her head was bowed in a posture suggesting either great despair or extreme fatigue. She raised her head as they approached her, and her youthful face broke into a faint smile when she recognised Ben. She inhaled sharply, but any attempt to call out to him was halted by Merisi holding up a cautionary index finger to his lips.

Instead, she accepted Richards' warm embrace with exhausted gratitude, resting her head on his shoulder as he encircled her with his arms.

With not a word spoken, Ben inspected the strange, transparent wire that was both binding Grazia's hands and securing her to the plinth. He pulled a penknife from his pocket but was halted by Merisi's firm hand. "No, Dottore," said Merisi. "A blade will not cut these bonds." He produced a tool that looked to Richards like a miniature soldering iron – within seconds, Grazia was free, throwing her arms around Richards's neck.

"Oh! This is my friend, Merisi. Thanks to him, I have found you!" said Richards.

"Yes, Miss Rossini knows who I am," smiled Merisi. Without offering an explanation, he turned to Grazia, "We need to act quickly, my dear. Your captors. where are they, and how many are there?"

"There are four of them. A nasty little Chinese *buco di culo*, and three others. They are down in the cellar, over there," she whispered, nodding towards the opposite side of the atrium. "They appear to be looking for something. Then there are two more – big ones – roaming around the building."

"Well, we won't need to worry about those two," said Merisi, with reassuring confidence.

"Oh?" replied Grazia, with a look of surprise.

"No. I have dealt with them already!"

"So, what is it they were looking for?" asked Grazia.

"Ancient Roman wisdom!" replied Merisi, "and I'm very happy to say they will not find any down in that cellar! Clearly, GIATCOM's multi-billion-dollar intelligence and surveillance system is not what it could be!" Ben looked at Merisi quizzically. "In fact," Merisi continued, "if Feng had an ounce of wisdom, to go with the encyclopedic knowledge he has been programmed with, he would not have walked straight past the very thing he is looking for!"

The words appeared to further confuse the reunited couple, but both could both see Merisi staring smilingly, at a small, oblong object of marbled stone situated towards the edge of the atrium, just a few yards from the plinth to which Grazia had been tied.

"What is it?" she asked Merisi.

"It looks like a small tomb," suggested Richards, "maybe of a baby, or even a family pet."

"That's exactly what it is supposed to look like," replied Merisi, "but the Romans were very clever concealing things important to them. What we are looking at is something more akin to a 'strong box,' a container of precious objects and important documents. If I am not mistaken, that stone has kept safe things that are of immeasurable value to us."

With that, Merisi cautiously crept into the pale light of the atrium and knelt down beside the box. There was no visible seam or join that suggested that there might be a means of opening it, if indeed it were a box, but Merisi appeared to know exactly how to reveal the secrets inside. Expertly, he applied gentle pressure to one of the top corners of the box, and sure enough, the top third of the object slid to one side. Glancing across the atrium in the direction of the cellar, Merisi lifted out a wad of browned vellum, and then closed the box with the same care with which he had opened it. "These are what we need!" he said, holding up the vellum documents, and clearly elated at his discovery, he rejoined Ben and Grazia. "Now – follow me. And don't make a sound!"

Cautiously glancing around in all directions, Merisi led them away from the main atrium through what was once a much smaller, secondary atrium, and into what remained of a smaller room. "This was the kitchen!" he whispered, turning his head slightly so that the pair following could hear. At the end of this kitchen, Merisi turned to his right and stepped over some yellow & black striped, plastic ribbon that was clearly designed to cordon the area off to the public.

"Where is he going?" whispered Grazia to Ben, holding his hand tightly as they picked their way through the ruined house.

"I don't know – but we'd better stick close to him."

At that moment, Merisi turned to his left and began to walk down a narrow flight of stone steps. "We are going down into the wine cellar, my dear," replied Merisi, "the one that I hope our 'friends' don't know about. Tread carefully, as parts of some of the steps have crumbled away."

Once at the bottom of the flight of steps, they were suddenly engulfed in darkness, as the cellar to the kitchen was far beyond the reach of the light of the moon – until Merisi lit up the underground space with a small, disc-like device.

The pale-yellow light revealed beautiful ancient paintings, whose colours appeared to be as bright and fresh as they had been when they were painted. They depicted every-day, rural scenes, which told the story of simple folk working on the land, representing each phase of the viticulture of that time. Each scene was connected by a continuous and elaborate frieze that was in the form of a winding vine. Many of the scenes were, ironically, painted with Mount Vesuvius in the background, its sharp peak not yet sundered by the eruption about to devastate the city.

Richards marvelled at the beauty of what was, after all, a purely functional space, and Merisi too appeared to be awestruck gazing at the walls, holding the light above his head – but Richards soon became aware of the fact that the square room was in fact totally empty. What's more, having scanned every inch of the four walls, he could see no sign of a SATOR SQUARE, or anything that resembled a script of any kind.

"Merisi?" he said. "There's nothing here. Why have you brought us down here?"

"No, Dottore, this is not our destination," replied Merisi, and with a smile, he turned and made his way to the far corner of the cellar. Suddenly, both Merisi and the pale light guiding them disappeared. Ben and Grazia halted.

"Merisi?" exclaimed Richards, as loudly as he dared. Then the yellow light reappeared, revealing the silhouette of Merisi's head, and the sharp edge of the wall around which he peered.

"Come! It's through here," said Merisi. He was inviting them to follow him through a narrow gap at the back of the cellar, which would have been invisible to all but those who knew of its existence.

Once through the narrow aperture, Merisi increased the brightness of the light.

"I think we should be safe here," he declared in a tone that was a little louder than a whisper. Ben and Grazia looked around a much smaller space with a distinctly lower ceiling.

This time there was no sign of the fine art that adorned the walls of the wine cellar. Instead, there were more informal images drawn by largely unskilled hands, which were unmistakably Christian in nature. There were depictions of Christ's crucifixion accompanied by more abstract crucifixes, on all four walls, which provided clear evidence of its former purpose. There was also at the base of the wall opposite the narrow entrance, the now familiar and remarkably well-preserved letters of a SATOR SQUARE, etched into the sandy stone render.

Grazia was the first to notice the design. "Look!" she cried, pointing with the index finger of her right hand. "It's one of those squares!"

Richards immediately felt a sense of achievement, a sense of arriving at journey's end, with the discovery of the Square.

"Yes, my dear," said Merisi. "It is a SATOR SQUARE – but more of that later," he added, glancing across at Richards. "First," he continued, "let us examine the contents of the stone box!"

Merisi knelt on one knee and spread the vellum documents on the dusty floor. Richards leant forward to examine what were clearly very detailed and highly technical drawings – but Grazia was already in front of him, squatting close to the ground, eagerly poring over the drawings. The accompanying writing in Latin script of some kind meant nothing to Richards, but from what he could see of the illustrations they were clearly penned by an engineer. They resonated with him immediately.

"What are these?" he asked in wonder.

"They are plans, Dottore," replied Merisi. "Construction plans by engineers just like you – except that they developed their ideas some two thousand years ago."

Merisi picked up a sheet that appeared to serve as the title page, for it contained just two words. "'Aqua Traiana,'" he read out loud, but in a hushed tone, as if telling a ghost story.

"What does that mean?" asked Grazia.

"It is a 1st Century Roman aqueduct built by the Emperor Trajan," replied Merisi, "and, look! It says on the next sheet it was inaugurated on the 24th of June, in the year 109 AD."

"It fed water mills on the Janiculum," said Grazia.

The two men looked at each other in amazement.

"So you know your Italian history, my dear," smiled Merisi.

"Yes – I know my history and political science."

"So, these must have been the original concept plans for the aqueduct, which were then lost when Vesuvius erupted in 79 AD," suggested Richards.

"Yes, and they show wind turbines being used on the top of the aqueducts," cried Grazia.

Richards now realised the full significance of the drawings. He knelt down to examine what was written, in smaller characters, in the bottom margin of one of the sheets. "Good God, man!" he exclaimed. "This is the work of Vitruvius!"

"Who is he?" Grazia asked.

"Marcus Vitruvius Pollio, commonly known as Vitruvius, Vitruvi, or Vitruvio was a Roman author, architect, civil engineer and military engineer during the 1st Century BC," replied Richards, slipping effortlessly into lecturing mode. "He was known for his multi-volume work entitled *De architectura*."

"A legend then?" Merisi enquired.

"Yes, he was one of the first great pioneers in Roman construction engineering and design," explained Richards. "It appears these design engineering concept drawings were 'work-in-progress,' and they were intended to be used. Vitruvius was born well over a century before Vesuvius erupted, but it's clear some of his advancements in design engineering which had been lost in the eruption had yet to be tried and tested."

"And I certainly don't know of any aqueducts, past or present, using this type of technology," added Grazia.

"Don't you know what this means?" Ben cried.

"Yes, it means the ancient Romans were well ahead of their time," suggested Merisi, "and who knows what they would have gone on to achieve by understanding how to harness the wind and use it to power their irrigation and sanitation?"

"It also confirms that wind power has been under-used as a potential resource, and this will continue to be the case, until we fully understand how it can be harnessed most efficiently, using the architecture of the landscape and the built environment," said Richards. "We need to learn how to use it to support other technologies."

They fell silent, trying to comprehend what they were looking at. Merisi slowly and reverently sifted through the sheets, browned with age, but which nevertheless presented sharp, clear drawings.

"This," said Richards at length, "is the oldest set of design engineering drawings I have ever seen, which show two different types of integrated renewable technologies working together - Wind and Water. And it is quite remarkable, and astonishing at the same time, to think these are the first signs of how hydropower could have been developed from Vitruvius's early ideas. Who knows? The world might have advanced to make greater use of wind and water turbine energy, and a lot earlier, and so make a lot more widespread use of it today."

"But how did the Romans get water up the hills to the aqueducts?" asked Grazia.

"Look closely at these drawings," replied Ben.

Richards lay the drawing out in front of them as he began to make notes on the drawing in pencil to help make sense of the Roman invention.

"The Romans were able to move small amounts of water uphill at great speed, and continuously in some situations using a water wheel powered by water flowing down one side of the valley. The wheel turns due to the power of the water flow and lifts the water up in paddles whilst the wind vanes rotate and push the water wheel with their connecting blades. Once the water reaches the top of the wheel, it is released and falls into a trough. It then drops into the aqueduct over much smaller water wheels, which are rotated by not just the falling water, but also the tips of the smaller wind vanes' blades. These small wind vanes help to push and rotate the smaller water wheels in the prevailing wind direction. In this case, the wind is blowing from North West to South East, so that the water flows along a slightly inclined aqueduct which runs towards the city of Rome."

"And what about this other drawing, Ben?" asked Grazia, taking a particular interest in a vellum sheet that was next to Merisi's knee. "Could this be an alternative option?" Richards proceeds to lay out the other drawing, whilst continuing to write notes on the drawing in pencil to help make sense of it...

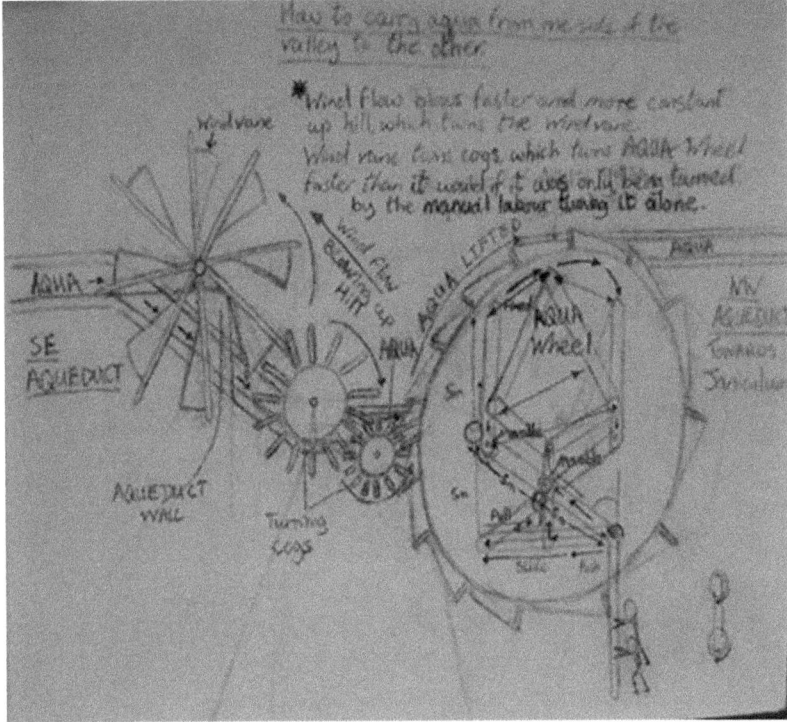

"Well, in this case," began Richards, "a wind turbine rotates at high speed with wind flowing faster and with a more constant flow up the hill. As the wind vane turns kinetic energy turns the cogs to power the water wheel, which lifts the water on each slat after it has flowed fast down the hill. The water is pushed down the opposite side of the valley continuously to enable the water to flow along the aqueduct to the Roman city."

"Because the wind vane is turning the cogs, the water wheel turns faster than it would if it were only being turned using manual labour alone. When there is strong demand for water use from the city, both wind power and manual power can be used to lift the water quicker and in larger quantities. There will also be a seasonal effect, as when it is less windy, more manual labour will be used to rotate the water wheel with less reliance on the wind vane rotating."

"You are starting to educate me, Dottore," said Merisi with a smile. "I understand now. Wind power can be used in conjunction with waterpower and converted into mechanical power, to increase the efficiency and speed of the rotating water wheels."

"That's exactly right," replied Richards.

Merisi paused and surveyed all the documents for a moment.

"I will tell you something," he said. "GIATCOM would give a great deal to get their hands on these drawings!"

"GIATCOM... who are they, exactly?" asked Grazia.

"A private advanced technologies company which deals with nuclear energy, cybernetic organisms, robotics and shape shifting technologies, on a global scale," replied Merisi. "In the future, they will spawn an evil culture throughout their organisation which will destroy society, and the landscape of the world through greed and corruption."

"So what would such an organisation want with me?" asked Grazia.

"I suspect they wanted to hold you as a form of ransom in order to force Ben to give himself up. His designs and knowledge are indispensable for them. They want to use them to help avoid the disasters they have caused – whilst not affecting shareholders' dividends, probably!"

Ben Richards himself was still avidly poring over the details in the ancient documents: "You can see they have even carried out an analysis of projected wind yield performance," he exclaimed, "and have introduced the concept of presenting their findings in the form of data. I can use this data to show how generically over generations of time there has not been much of a difference in wind speed and yield across a flat surface, such as a long-span roof. This is ground-breaking stuff - and shows how far advanced the Romans were in science and mathematics, as well as in architecture."

"The Romans' data is similar to my own wind tunnel test results. It backs up the idea of using the architecture of a surface to yield a more constant flow of wind along it, than in comparison to large freestanding turbines. As you can see clearly from their drawings, their plans were to introduce smaller wind turbines on the roofs of amphitheatres. In comparison, my studies showed a two-megawatt freestanding option versus two options of roof mounted application; 86 x six-kilowatt versus 120 x six-kilowatt micro turbines."

"My results show that small roof mounted turbines performed at more than twice the efficiency of a two-megawatt turbine power output. The 86 x six-kilowatt roof mounted turbine configuration has an efficiency of 39.4% as opposed to the much lower two-megawatt freestanding turbine having an efficiency of 17.5%, which is a significant improvement in energy generation! It means if you were to have a roof mounted configuration of the same megawatt input size of a two-megawatt freestanding turbine, then a two-megawatt roof mounted turbine configuration would generate almost one-and-a-half times more energy output than a two megawatt freestanding turbine array would. That's a 150% increase in energy yield performance, which is a significant increase! This could be really significant if rolled out globally on stadia on a large scale."

It had dawned on Richards that these findings suggested a more constant flow of the wind passed through the roof mounted turbine configuration, as opposed to the free-standing turbine. The Stadium Roof Mounted Turbine Configuration was using the Coanda effect to enhance the efficiency of the rotors, enabling them to turn more consistently, thus creating a more constant wind flow. This was all based on using the curved architecture of the roof to help avoid the occurrence of wind turbulence and wind shear.

"This suggests that the best value option based on the wind yield generated from the amount of capital invested is the roof mounted option," declared Richards, "due to the turbines' rotors generating wind yield more efficiently. This is assuming one can prove that a two-megawatt freestanding turbine is much more expensive to invest in than a roof mounted turbine array of similar megawatt output."

"Indeed, Dottore. And the Romans also built this beautiful city that we see around us," said Merisi, "but we must now turn our attention to more pressing matters."

Richards knew exactly what his companion was referring to and stared vacantly into Merisi's eyes. "I have a journey to make, don't I?" said Richards.

Merisi sensed reluctance in the young engineer's voice – a certain apprehension. "So, why the reservations? I thought that you would be excited by the prospect of visiting the future. After all, you will be the first person from your time to do so. Do you fear venturing into the unknown?"

"I wouldn't say 'fear' – but please understand that you are asking me to do something that my every fibre knows to be impossible."

"'It always seems impossible, until it is done' – Nelson Mandela," retorted Merisi.

"Yes, yes. I understand that – but how?"

Merisi fell silent for a moment as he pondered over how much information he should divulge to the eager young scientist.

"Is it done through wormholes? Anomalies? What?" demanded Richards, impatient for answers.

"Well," began Merisi, at last, "it transpired that science fiction writers and scientific theorists were never far from the truth. Time can be bent and shaped. It has properties very much like space, matter and antimatter. The problem with time was with the vast amounts of energy that were required to alter it. The irony, Dottore, is the course of time can be deflected in much the same way as the vertical trajectory of a stream of water can be changed."

"The Coanda Effect!" said the two men in unison. Merisi nodded.

"As you described so often in your presentations, in the absence of another force, gravity will cause water to fall vertically, but put a curved surface next to that stream of water, and the effect of friction and drag will cause the water droplets coming into contact with that surface to adhere to it, and so change direction. Time behaves in exactly the same way, if enough energy is applied at a given point in time and space in order to create that friction."

"And what was that great leap forward?" asked Grazia, appearing to be every bit as fascinated by the science as Richards.

"Forward?" snorted Merisi, unable to hide his disdain.

"Mankind made a leap into the darkness - and leapt 'forward' to The Stone Age, in most parts of the world." He immediately detected a look of horror on Grazia's face. "Apologies, my dear. How could you have known? It is a fair question."

Merisi ran a sun-tanned hand through his thick, black hair.

"It was GIATCOM, with their laboratories based in China that made the crucial breakthrough. Are you both familiar with the process of creating energy through nuclear fission?"

"It's not my field, but I think I understand the principle," replied Richards. "It's something to do with the energy released when an atom is split into two constituent parts. Isn't it about releasing the attractive energy that binds the constituent parts of the atom together?"

"Yes, that's the principle," replied Merisi. He then engaged Richards' attention with a fixed stare. "Imagine the energy potential that can be released by breaking protons and neutrons into a hundred yet smaller parts!"

"Wow!" exclaimed Grazia.

"...and, very soon, GIATCOM's scientists were able to control this reaction and harness the energy that was released. Suddenly, the knowledge of this process was 'the new oil'. What was once a large corporation, part financed by the Chinese Government, and now became the dominant superpower on earth. Its Board of Directors could now control the world's markets, and after the inevitable weaponisation of the technology, its politics as well."

"For how long was this discovery kept a secret?" asked Ben.

"For several decades. China was never a difficult place for the keeping of secrets, and research and development was able to proceed unchecked and unrestrained in GIATCOM's underground laboratories in the Taklamakan Desert. Of course, the Japanese were the first to know about this – and disbanded their entire armed forces apparatus almost immediately! The world applauded this sudden move as a unilateral gesture of peace – and awarded the Emperor of Japan the Nobel Peace Prize that year."

"But someone must have been suspicious. Surely!" suggested Grazia.

"Yes, of course. But GIATCOM had such a head start, by this

time. Also, with China's borders secure and its enemies effectively neutralised, the Chinese Government was able to combine the full weight of its science and technology sector with GIATCOM's scientists, in order to realise the ultimate dream – to alter time, and then eventually, to master it."

"It took several decades in order to achieve it, but through a combination of advanced espionage and colossal investment in research, the United States of America with the support of the other advanced nations of the Western world developed their own programmes. The western world began to catch up with the power that had risen in the east. Once GIATCOM got wind of some early successes the Americans had with what was termed 'time warping', conflict was inevitable. The catastrophic results of this conflict you will soon be able to see for yourself, Dottore."

Ben and Grazia said nothing – and so Merisi continued:

"In the decades that followed the Global Nuclear Engagement, there has been a Global Alliance presided over by the UNA - the organisation that sent me here. But it is too late now. Over half the earth's surface has been turned into a nuclear wasteland, and what remains habitable is grossly overcrowded. It is only a matter of time."

"What exactly is the UNA?" asked Grazia.

"The 'United Nations Authority.' GIATCOM is now a full member of the Alliance, along with most of the remaining nations of the world – although many still do not trust their motives, and none of us know what weapons and technologies they still hide in their Taklamakan strongholds. One thing we are sure of however is that GIATCOM still has the capability of manipulating time in exactly the same way as the UNA, in order to maintain its own agenda. And the presence of Shui Feng here in 21st Century Italy proves that."

"The cruel Chinese one!" said Grazia, coldly. "And what do you know of this - Feng – exactly?" she asked.

"A very advanced, and very dangerous android," replied Merisi.

"Do you think he was in Milan at the auditorium?" asked Richards, beginning to understand at last, that there might be an explanation for the two near misses that he had experienced with the darts, both during the seminar and at the Ponte Vecchio.

"I am absolutely sure of it, Dottore – but he has the ability to change his appearance, and to do so very quickly and very convincingly. I have been on the look-out for him since I saw him arrive through the portal in Siena, and I've not seen him since – but that does not mean he has not been close by."

"And what do you know of his mission?" asked Grazia.

"Well, in general terms, to foil me in my mission."

"Which is...?" she asked.

Merisi paused. "In my time," he continued, "we have an expression about 'letting the genie out of the bottle.'"

"Yes, we have that expression here – I mean 'now,'" replied Richards.

"So, you will understand, Dottore, what I mean when I tell you it is my task to make sure that GIATCOM's genie never comes out of that bottle. In fact, what I must do is to break that bottle, to ensure the world has a future driven by wind power and other sustainable sources of energy. If the nuclear industry is allowed to survive and flourish, then quite simply, the world's future is bleak, as I have seen."

"And you are sure that Ben can make the difference?" asked Grazia.

"Not just me, my dear. The UNA has had teams of historians and statisticians working on the issue for many months, and their conclusion was – will be - the turning point in the world's history – the point at which the domination of nuclear power became inevitable – was Dottore Richards' assassination at the Milan seminar."

Merisi could see Richards was still having some difficulty digesting this notion, so he waited for him to regain his composure.

"Obviously, and happily, that event has been avoided," continued Merisi, "so history has already begun to rewrite itself, but there remains much to do. You still need to follow through with your ideas. You still need to find a way to make turbine-generated power in urban areas succeed, both technically and economically. It is with that conundrum I can't help you."

"Of course, there are fatalists – and fatalism has become very

popular where I have come from – who believe the general course
of history cannot be changed. They believe the nuclear holocaust
will eventually come, in one way or another, and they would tell
you, if they were here, your premature death is also unavoidable,
and you will be murdered, if not today, then tomorrow, or the
day after."

"And what do you believe?" asked Richards.

"I believe fatalism is nonsense – and this is why I am here.
To believe in fate is to live without hope – and how can anyone
live without hope? I am sure we are in control of our own destiny.
You have to believe this Dottore, as you will now be responsible
for the fate of us all."

There was another long pause, as Ben Richards struggled to
process everything Merisi had told him. He took hold of Grazia's
hand, gripping it tightly. Was it possible his ideas about wind-
generated energy could be the key to the world's future? He had
never doubted he had an important contribution to make to
the industry and had always had a strong conviction that wind
turbine technology was the way forward for the world's energy
needs. So why was he feeling so much self-doubt, now that a man
had travelled from the future to confirm these things?

Richards was the next to speak.

"And your experts are absolutely sure I am the person who
can reverse the world's fortunes?"

"There was no doubt as to their recommendations – and the
decision to act upon their recommendations was ratified by the
Executive Committee of the UNA, which is the very highest level
of authority."

"Please try to understand I have not been sent here on a
pleasure trip. Attempting to change the course of history has
always been considered being a response to the utmost need, and
the final resort. The year 2112 has seen a marked worsening in
the civil unrests, mass migrations caused by famine and disease,
and the near complete collapse of law and order in the remaining
cities. It is indeed ironic having mastery over time, which has
been what men have dreamt of for centuries, is now just about
the last thing mankind has control of – and it is on this we pin
our final hope."

Richards took a long, deep breath, and sighed. "Well, assuming this is not all a nightmare, and you are not telling me a pack of very imaginative lies – what do you want me to do?"

"First, Dottore, you need to go to the future to see for yourself. That will remove any trace of healthy scepticism you might have."

"...and you have the means to send me there, do you?"

"Well, it will not be me. I cannot do anything from here. A time portal can only be opened from one side – the side where the technology exists to make this possible. The time and location for the portal opening has been pre-set. My task will be to send you through the portal at the pre-determined time. You are expected, so a delegation of UNA officials will be there to greet you and give you a full briefing."

Then Merisi turned to Grazia and said: "They were not expecting to receive a second person, but I am sure you will be made very welcome, my dear."

Grazia looked at Ben as if seeking reassurance, but the engineer said nothing.

"I think it would be a good idea, Dottore," suggested Merisi. "Please understand, Miss Rossini," he said, then turning to Grazia, "Dottore Richards has grown accustomed to doing things on his own – but I am sure he will benefit from the support of 'his other half."

Both Ben and Grazia were taken aback and slightly embarrassed by this suggestion.

"Besides," continued Merisi, "with Shui Feng and what is left of his team wandering around out there, I think the year 2112 will be the safest place for you, my dear." Grazia smiled, asking no further questions.

"I will stay behind and await your return. I need to be here to close, or at least obscure the portal, to make sure nobody follows you on your journey."

Richards began to perspire as the reality of what he was being asked to do started to sink in.

"Will we feel anything, as we pass through the portal?" he asked.

"Not a thing. It is slightly counter-intuitive, I know, given the extreme levels of energy being expended in order to make your journey possible – but it is just like walking through a door. Although I'd be prepared for the climate in the 22nd Century being rather hotter and more humid than you are both used to!"

"All you need to do, when the portal opens, is walk towards the SATOR SQUARE, and step over the threshold."

"...and when is the portal supposed to open?"

Merisi consulted his hand-held device and smiled. "In precisely 40 seconds, Dottore. I have been keeping an eye on the time!"

Ben and Grazia both turned to face the SATOR SQUARE. "So, technically – if this all works – we will become the very first human beings to travel through time, given your journey won't actually take place for another 95 years," he suggested to Merisi.

"Well, that's not strictly true – but let's not go there!" replied Merisi. "What we must hope though, is you are the LAST human beings to travel through time, because if everything goes to plan, our actions will mean there will be no GIATCOM, no massive leap forward for nuclear science, and so no time travel!"

Richards turned his head to look at Merisi. "But that means that, for you, this has very much been a one-way trip!"

Richards detected the beginnings of a smile on Merisi's face as he looked away and sought to focus his attention on the SATOR SQUARE. He caught sight of a verse from the Bible that had been etched into the wall:

"John 3:16 For God loved the world so much that he gave his one and only Son, so that everyone who believes in him not perish but have eternal life."

Then the darkness of the room was transformed into a fluorescent blue light, and Dr Ben Richards and Grazia Rossini, hand in hand, stepped forward, into the future.

CHAPTER 12

THE DISAPPEARANCE OF DR RICHARDS

Three weeks later...

Tony Clarke paced across the floor of his spacious, 9th floor, London office, ignoring the magnificent view of Shard rising above London Bridge Railway Station, nearby. Clarke tried numerous times to reach Dr Ben Richards on the phone number Richards gave him when the two met in Milan. The phone number had been unobtainable since Clarke's return to England. The entrepreneur managed to speak to Ben's Personal Assistant - Phoebe Lomax and learned that nothing had been heard from him since the conference that both men had attended.

"It's not all that unusual for him to go off on his own when he's on business; he can sometimes be a bit obsessive about his work," Phoebe explained, "We are worried about him, he's never been out of touch for so long. So I called the police this morning to report him missing."

Clarke could tell from the slight but noticeable wavering of her voice the concern was genuine, and so he began to fear for the young man.

"Unbelievable!" he muttered out loud to himself. "How life can change! On top of your game, presenting to an international audience one minute, then missing the next." Clarke stared out of the window for a few moments and felt a tingle run down his spine, then a feeling of real sadness as he recalled the moment when he realised he shared a true vision with the young man. He felt he owed it to Richards to try to pursue that vision, although he could not escape the feeling that it was the young engineer who had the vision. The visionary leadership qualities to be the real game changer when it came to the development of creating original and innovative energy generation.

"Maybe it's a fat, middle-aged entrepreneur who has to take up the baton, and change the game!" he thought to himself, as he scanned the skyline of Central London. He imagined a future landscape where every tall building had its own version of a wind-driven power generator. "Maybe I'm the one – the captain of industry - who can finish what the engineer has started."

CHAPTER 13

MIRROR IMAGE

As the fluorescent blue light that briefly engulfed all his senses receded, Dr Ben Richards became aware he was standing in the centre of a large circular rather sterile room – a laboratory, he thought – but one that contained very little equipment. The room was dimly lit with the main source of light appearing to come from downlights positioned between the roof and the ceiling, for the full perimeter of what he could see of the room, without turning around. And Merisi was right. Travelling through time – if that is what he had done – had been surprisingly uneventful.

The first thing Richards was aware of was the warmth and softness of Grazia's hand, which was holding onto his with a firm, but child-like grip. Then, as his eyes adjusted to the light, he noticed a group of four people gathered informally, just a few yards in front of him. Two men and two women, all bespectacled and dressed in what appeared to be a one-piece uniform. Grazia let go of Ben's hand as the nearest of the four approached. He was a man with a round, unmistakably Chinese face, probably in his mid-thirties, and with a scant trace of two days' worth of facial stubble on his chin and upper lip. He smiled as he extended a welcoming hand to Richards.

"Dr Richards!" he said. "Welcome to the 22nd Century!"

Richards noticed that the man's uniform was adorned with a badge that had 'UNA' in its centre, as he shook his hand – but smiled, and said nothing. The Chinese man then turned to Grazia, with a quizzical look that made it obvious he had not expected to see her.

"My name is Grazia Rossini," she volunteered. "I am a companion of Dr Richards."

"She is a journalist and has been writing an article on my work," added Richards.

"Well, we were not expecting a second person to arrive," said the man, slightly uncertainly, "but you are both most welcome." He paused briefly, before continuing in a more formal tone of voice: "Allow me to introduce myself, Dr Richards. My name is Colonel Qiao Liang. I am responsible for coordinating the mission to prevent your assassination – or rather your former assassination - and transport you here today. I report directly to General Paul Steyning, who has overall authority for the operation. I am sure everything that you have experienced, and all you are about to witness, will seem totally counter-intuitive to you – but I can tell you, you are in the Time Portal Generation Room of the UNA's Beijing Base. It is 2.40am on the 7th of February 2112. How do you feel?"

Richards shrugged. He felt fine, physically, but had so many questions to ask about the new world into which he and Grazia had just stepped into. He didn't know where to start. Colonel Liang appeared to sympathise with Richards' initial bewilderment.

"It will obviously take a little while for you to get used to your new surroundings. We are currently four stories underground – and that is not a bad thing at this stage, as the real shock that awaits you will come when you witness Beijing at ground level!"

"But first," he continued, "We need to show you to where you will be spending the remainder of your first night here. I am sure you will both be very tired. A full debrief can wait until the morning." At that moment, the colonel noticed the beige sheets of velum peeking out from within Ben Richards' jacket. "Ah! They must be the drawings!" he exclaimed.

Richards had almost forgotten he had stashed them there just prior to facing up to the portal. "Yes," he said. "Merisi appeared to know exactly where to find them."

"He did, indeed," replied Colonel Liang. "Merisi is an excellent operative, but even he is but the tip of an iceberg. The Agency dedicated a great deal of time and resource in locating the drawings." He extended a hand, as if inviting Richards to relieve himself of the burden of such valuable artefacts, but Ben felt a strange reluctance to give them up.

"Would it be possible for me to hold onto these until the morning?" Liang asked. "There is so much that I want to know about the Romans' ideas and approach."

Richards appeared doubtful. "You must understand, Dr Richards, that the drawings you have in your possession are absolutely priceless. They are of immense value to us."

"Of course."

"Recent – let us say 'indiscretions' – have not only destroyed much of modern civilisation but also wiped out most of what remained of the fabric of previous civilisations. Unfortunately, I have to tell you that nothing now remains of the ancient ruins of Pompeii you have just left – and the city of Rome, you will know very well, also suffered the same fate as London and Paris."

Liang ended his account abruptly. He could see the revelations of the fate of the world that Ben Richards knew were very shocking for the young man – who remained silent. The pause enabled the colonel to reconsider his position, and to reflect on the importance of his very special guest.

"Ok, Dr Richards," he continued. "You may keep the drawings until morning, but please let us take copies first." He beckoned to one of the people standing behind him – a woman in her mid-twenties, who Ben could tell from her less elaborate uniform, was of a considerably lower rank. She handed the colonel what appeared to Richards, a thick, plastic, transparent folder. Liang invited the young engineer to place the documents in the folder – which he did as one wad. A bright, white light emanated from the folder as soon as it was closed.

After no more than a couple of seconds, Colonel Liang Opened the folder and returned the documents to Richards.

"We have a full set of copies now," he said, with a smile,

"but please look after those originals. We have gone to a great deal of trouble, and they are very precious to us."

Richards nodded his agreement, and very carefully placed the velum artefacts within his jacket.

"Now – is there anything else either of you requires?" asked Liang, running his eyes over Grazia, who was the more dishevelled of the two. For the first time, Richards realised she was still dressed in the clothes she was wearing when they met in Caravaggio's restaurant – although her red blouse was rather dustier than it had been that evening.

Grazia could sense she had suddenly become the centre of attention in the room. "I'm fine," she answered instinctively, "Although a change of clothing would be good!"

Liang smiled. "Yes, clothes will be provided for you both. I cannot guarantee they will match the high standards of fashion of the era you have just left – but they will help you blend in."

"Now, I will leave you in the hands of my colleagues, who will take you to your overnight quarters." Colonel Liang then introduced each of the three people standing close by, but Richards was, by this time, too tired and mentally fatigued to take in who they were. All he wanted to do was to get some rest and have a closer look at the engineering drawings the colonel had entrusted him with. He was also keen to spend some time with Grazia and catch up with what had happened to her since that dramatic night outside the restaurant. He looked across at her and was rather surprised to see that she looked surprisingly calm and unruffled.

As they turned and began to follow Liang's colleagues out of the room, Richards reached out and touched Grazia's hand, gently taking hold of the little finger and ring finger of her left hand. "Grazia – are you okay?" he asked.

Grazia appeared slightly surprised at the question being posed. "Yes. Yes, of course," she replied, taking her hand away from Ben's as she used her fingers to brush her lank hair off her face, tucking it behind her ear. "Yes, I'm fine. I'm just very tired. That's all."

"Before you go," called Colonel Liang, in a manner that was loud enough and authoritative enough to stop all five immediately, "I should inform you, that you will undoubtedly feel a heavy security presence during your stay. That is because we are expecting a visit from President Milton Westcroft himself, tomorrow. He is flying in from our headquarters in Melbourne – and he is very much looking forward to meeting you, Dr Richards."

With that, Ben and Grazia followed the UNA officers into a spacious lift where, as if to support Liang's warning about the current security presence, were joined by four more officers, whose size and bearing meant there was no mistaking their role as being that of security. Richards observed them closely as the lift descended deeper into the bowels of the UNA establishment. They betrayed no emotion, and they were probably 20% larger in all respects than any human that he had ever seen, and so he wondered just how human they might be if they were human at all. They carried no visible weapon – but their streamlined uniform included all manner of compartments and holsters in which weapons might be hidden. Richards concluded that he was grateful that these large units were there to protect them!

The slight figure of Grazia was hidden behind one of the guards during the short trip in the lift, and so the next time Richards was able to see her was when the lift doors opened, and the guards exited the lift and marched across to the opposite side of a wide corridor. They were soon followed by the three officers leading Ben and Grazia along a curving corridor with many doors, all of them identical in the manner of any number of 'chain' hotels that Ben had experienced during his career of conference and seminar presentations. Suddenly, the female officer who had been entrusted with leading the small party stopped opposite a door numbered '825.' She turned to Grazia, extending an arm to pass her a thin, plastic bracelet, said: "This is your room, Miss Rossini. This contactless bracelet will lock and unlock all the doors, you can use it to communicate with us, should you need anything."

Grazia thanked the officer as she took the bracelet, and then used it to open the door to Room 825. Then, almost as an after-thought, she turned to Richards and said: "Good night, Ben. I will see you in the morning," and with a faint smile slipped into the room and closed the door behind her.

* * * * *

Once inside the room, Grazia stood by the door and listened. She heard the female officer speaking to Richards just a few yards away, before hearing the sound of the door in the next room opening. Richards was trying to establish the timing of the following day's activities – apparently there was to be a full debriefing mid-morning, over a late breakfast, before an afternoon meeting with President Westcroft and his senior aides at the UNA's office in Central Beijing.

There was a formal exchange of thanks and 'Good Nights, before Grazia heard the door of Ben's room close.

Grazia peered through the wide-angle peephole in the door and saw the delegation of UNA officers hurry past back in the direction of the lift. She could also see the shoulder of one of the large guards that had accompanied them to their room – and on the other side of the peephole, there was the figure of a second guard, silhouetted against the main light of the corridor.

The room was more akin to a spacious apartment, with a kitchen area opening out to a luxurious lounge, and with a sliding door leading to a bedroom and en suite bathroom. But Grazia marched straight across the space and sat on a low sofa that backed onto a large window. She removed a small, disc-shaped communication device from her trouser pocket, ran her palm over the surface and held it up to her mouth.

"Cardus?" she demanded. "It's me... Yes, I am here, in Beijing... Yes, Richards is here – and he has the drawings... Merisi? No, he is still alive. He didn't travel back with us... No, no trouble at all. Merisi is a fool. Not only did he lead me to the artefacts, he virtually gave them to me in my hand! No, I didn't, but I must say that it was very tempting – but I can deal with Merisi later... It appears that we are in a secure UNA military establishment, quite a long way underground... No, don't try to send a team. We are being heavily guarded. Besides, there is no point, I am on the inside now. The plan is for us to meet with Westcroft tomorrow afternoon, in the centre of Beijing, just around the corner from the main GIATCOM office. I will deliver the drawings to your people before the sun sets. Tell them to expect me."

Without waiting for a response, Shui Feng extinguished the faint light on his device and placed it on the small table beside the sofa. He then laid his head on a small, beige cushion, and drew his legs up to his chest. And there he remained, adopting the pose of a large sleeping cat – but with his eyes wide open, staring straight ahead, biding his time until dawn.

CHAPTER 14

ARRIVAL AT THE CITY OF A MILLION WINDOWS

The UNA transporter – a craft with a perfect aerodynamic shape resembling a large, flat beach pebble – glided effortlessly through the permanent fog shrouding Greater Beijing. The sandy- coloured air, which had been made toxic by heavy industry many decades ago, was now dangerously radioactive and lethal to any man or beast lacking the high-tech protection developed rapidly in what was left of the inhabited world. The transporter, a Douglas-Wisson 13-35, in common with all other aircraft of the era, was equipped with such coatings and air filtering equipment necessary to keep its precious cargo safe from contamination. The craft was also large enough to comfortably accommodate the party of 38 personnel, most of whom travelled with a security brief, plus the required weaponry and emergency supplies necessary for this short, routine flight to the centre of the city.

Shui Feng sat close to Ben Richards – holding tightly and

attentively to the engineer's hand - in the centre of the carefully planned seating arrangement. Feng had used the overnight wait to review the events of the previous few days and had concluded that he might have been more convincing in the role of Grazia. He felt his performance might have been a shade detached, and so he had begun the day 'in character' by giving Richards a warm hug and a welcoming kiss when they met for breakfast.

Shui Feng's objective was very clear now. He had to avoid detection until the chance came to seize the drawings Merisi had virtually presented to him in Pompeii – and then he had to be decisive and ruthless. He had of course rehearsed in his high performing synthetic mind any number of contingency actions, should his shape-shifting disguise be discovered at an inopportune moment.

Feng's task was made easier by the fact that Richards, who was the person most likely to notice any 'un-Grazia-like' behaviour, was by now, almost completely absorbed in the ideas he was generating as a result of studying the drawings. He had managed just a couple of hours' sleep, having spent most of the night poring over the ancient velum artefacts. The fact that nothing but a sand-coloured haze was visible through the transparent sides and roof of the aircraft and the silent vibration of the aircraft had both lulled him to sleep, and he snoozed quietly.

Shui Feng had watched keenly as Richards handed the drawings to Colonel Liang, who joined them during breakfast, prior to the formal briefing session. Feng's eyes had then followed the colonel as he walked over to someone who appeared to be in charge of security – a human, he thought, yet with all the stiff authority of a security specialist. The type of a career soldier on whom generations of androids had probably been modelled. A soldier who had been entrusted with the documents' safekeeping, and their subsequent presentation to The President himself. Feng had constantly monitored this individual's movements during the morning and now directed his gaze so that he could observe him in the periphery of his vision; he was seated in the row behind Feng and Richards, and five seats to their left.

The briefing was short and to the point, and Feng had learnt little he did not already know or could guess . As expected, both very important guests of UNA were to be presented to President Westcroft, and apparently the old statesman, as a keen scholar of the 20th and 21st Centuries, had plenty of questions he wanted to ask them about the world they had just left behind. Of greatest concern to Feng was the fairly casual reference made to 'routine security checks' –he would be instantly exposed as being very much a product of the 22nd Century, so he resolved to make his move well before that part of the programme.

It was just past noon when the craft began the descent to the delegation's destination – but the sun remained invisible behind the smog. What changed, though, was that the City of Beijing, now universally known as The City of a Million Windows, now came into view beneath them.

"Look, Ben!" said Feng, nudging Richards firmly enough to ensure that he would awaken from his slumber. "The city!"

Richards opened his eyes with a jump and a grunt and was immediately in awe of the sight before him. His utter astonishment at the scale of the city had been expected by the other members of the party, and eyes and smiles were already turned in the direction of Dr Richards and Miss Rossini, to register their aghast reaction. Shui Feng did his best to appear surprised at what was for him, a very familiar sight – but Richards' awe was entirely genuine: "Well, I didn't expect this!" he muttered, as he unbuckled his seat belt and walked towards the large, continuous window at the front of the craft.

There were skyscrapers and other architectural structures built on a scale that seemed unimaginable to him, and instead of the predominantly vertical cityscape that he associated with 21st Century cities, he noticed horizontal development, using elaborate twists and curves, was just as prevalent. What added to the beauty and magnificence of Beijing was, with the sun being permanently obscured, artificial light was required around the clock on every day of the year, creating a dazzling panorama of light.

Between the buildings were what appeared to Ben Richards, to be ribbon-like structures. As he looked more closely, he realised that these were made up of many individual vehicles travelling through the air in closely packed convoys which were three vehicles deep, and he guessed about half a dozen vehicles wide. He could only wonder at the technology behind this strange form of transport – but noticed to his amazement, not one individual vehicle emitted its own light. This, he concluded, must be because the whole city was bathed in permanent light, but it also suggested all units within each convoy must be automatically propelled and guided. Because of the lack of headlights and taillights, the streams of traffic were quite hard to detect. They could only be seen against the background of the lights of the buildings, the more Richards studied the scene. More and more vehicle convoys he detected, as they snaked between and within the astounding architectural forms.

Richards gazed in silence at the changing scene through the clear glass of the craft until it came to rest on a landing stage, roughly halfway between the top of the tallest buildings and the ground far below.

"So – how many people actually live in this city?" he asked Colonel Liang, as he and Feng were ushered from the craft and onto the platform – where the air was evidently safe to breathe.

"People?" queried Liang. "It depends on how a person might be defined..." He could tell from the quizzical expression on the face of Richards, who was unaccustomed to a world that included enhanced humans, part-humans, and androids of various kinds. His response had not been helpful – so he changed tack: "- err, if you mean 'individuals,' then the population was just under 200 million, last year, but the peak population for the city was nearer 400 million, four years ago."

Richards was finding the scale of this new world very hard to take in, but as the party made its way across the landing platform, towards a number of smaller vehicles, he became aware of a sight appearing all too familiar. A short distance from where the craft had landed, and away to the right, was the tall, seemingly boundless building that he had seen in the dream he had in the car on the journey to Milan. Richards stopped. "I know that building!" he said, gazing up at the glass façade that reflected all the lights of the buildings around it.

Colonel Liang looked at Richards with an expression that betrayed some scepticism. "That is the headquarters of the

GIATCOM Company – how do you know it?"

"I saw it in a dream," replied Richards, "a very vivid dream." He turned to Grazia. "Don't you remember the dream I told you about? In the car when you were driving us to Milan?" Shui Feng fidgeted uneasily. This was one challenge that he had not anticipated and had no contingency action to hand. "Surely, you must remember," insisted Richards. "You woke me up, and then I told you all about the strange dream I had."

All eyes were now on Shui Feng – who decided upon a tactic of evasion: "Yes, but it was just a dream, right?"

"Yes," replied Richards, "but everything is just as it was in that dream – right down to advertising screen on the side of the building." He pointed to an enormous screen that was positioned adjacent to a landing platform, which was of a similar nature to the one that the party was standing on and positioned at a similar height.

On the large advertising screen was a large and illuminous word in the colour red, which read 'STEER.' The advertising screen then flipped to show a caricature image of himself graffitied on a wall with the words, "Different is good, Dr Richards."

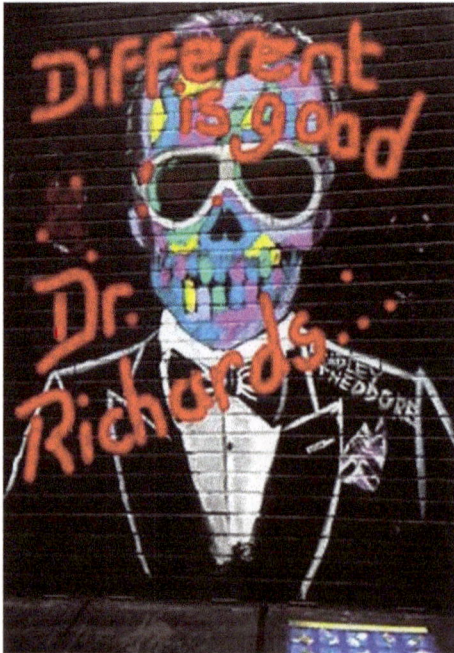

He realises the word 'STEER' is a coded anagram that Merisi told him to be aware of. He remembered being told he needed to be aware of random words that may appear in the future trying to get his attention, but not disclose them literally in fear of GIATCOM understanding what is being told to Dr Richards from UNA.

It took him a few seconds to decipher the anagram. He realised the word spells 'RESET' and immediately understood what it meant to him. The year 2024 post COVID-19 pandemic needed to be 'RESET' with rethinking and adopting his innovative renewable energy and environmental solutions, so the future can be saved.

The message 'RESET' is to tell Richards that, back in 2017 he needs to be prepared for the 'RESET' coming, and his vision and ideas are going to play a big part of him steering the 'RESET' in 2024, post the COVID-19 pandemic.

Richards then snapped out of his trance when the next advert came up on the screen. It showed a brief history of climate change and its link with airborne viruses over the past century. The screen showed the words: 'COVID-19 pandemic and its link with climate change.' The following words rolled down the screen whilst video footage of a series of natural disasters and pandemics from the 21st century played in the background:

Rethink Air Pollution and its link with airborne viruses!

The spread of viruses is made worse in environments with poor air quality.

A reduction in oxygen levels, increased CO2 emissions, and dense populations, can lead to breathing difficulties and lung diseases. During a pandemic such as the global Coronavirus crisis in 2019, people with already impacted health found their health exacerbated due to the poor air quality in the areas where they lived. They are especially at a high risk of further infection from such virulent viruses in locations where air quality is poor, and populations are dense. It is each and everyone's responsibility to recognise that there is an oxygen debt in poorer air quality locations, especially polluted air within densely populated locations around the world. We all should rethink the ways by which we can increase oxygen levels, improve air quality, and support the health and wellbeing of our family and friends. Important clues in the patterns the disease is leaving that tell us quite a bit about what conditions can hasten its spread and even worsen its lethality. The areas of the world where most people died of airborne virus diseases with the highest fatality rate were densely populated cities.

Air pollution was blindingly severe in the region of northern Italy where the coronavirus was firstly most virulent. The Lombardy region and the Po Valley in northern Italy ranked among the most air polluted areas of Europe, at the time of COVID-19 pandemic.

All this dirty air does not just clog vistas, it also clogs lungs and the respiratory system. Basically, tiny pollutants 50 times smaller than a human hair can enter the lung and sometimes get into the bloodstream , compromising the immune system.

Without normal healthy mucosa, the nose and lung lose the ability to slough off bacteria and viruses typically inhaled.

Lungs normally clear pollutants through the removal of viruses and bacteria by coughing. Healthy nose hairs also block the inhalation of pollutants. But chronic air pollution compromises the ability of the lungs to do their job. The natural mucociliary escalator dries up and cannot do its job of keeping us healthy.

The Coronaviruses that came after COVID-19 were a continued threat and was greater for polluted cities. Clean air returned to cities in lockdown. But it was short-lived after the lockdown. As society was not able to make the complete paradigm shift to working in a more dispersed rural way to reduce the amount of fuel emitted to the atmosphere within cities straight away. It wasn't until 2040, when all modes of transport were powered by renewable electrification and working from home permanently became our place of work. Buildings had limited occupancy. But the damage had already been done.

Air pollution continued to cause hypertension, diabetes, and respiratory diseases due to continued poor air quality, these were all conditions which doctors linked to higher mortality rates caused by COVID-19 affecting all these conditions.

By lowering air pollution levels or extracting the pollutants from the air, we can help the most vulnerable in our fight against this and any further future pandemics.

Richards nods and sighs and thinks we haven't advanced with climate change and renewable technologies enough in 100 years on from 2017. I need to design an air extraction device that takes viruses and high CO_2 emissions from a given air space to improve air quality.

-He was then interrupted by colonel, "Shall we go Doctor?"

"Yes certainly," and then to Richards' surprise the next advertisement looked familiar. "No, wait colonel. I want to see this."

To the engineer's continued amazement, he was also to remember the precise advert that ran across this screen that followed. It showed a video of a small, mauve vehicle darted between cloud-high buildings, and then plunged beneath a turquoise sea and propelled itself, like a torpedo, around underwater atolls and coral gardens. Just above this screen, he could see that the top of the towering edifice was designed like an ancient ziggurat, with steps ascending towards a pyramid-like summit within a shimmering glass dome. At the tip of the pyramid, and sitting on the outside of the dome, there was a giant hammer with lightning bolts shooting from both ends of its head – the logo of the world-dominating GIATCOM Corporation.

Richards was then astonished, not to say horrified, to see the scene of his dream being played out in front of his very eyes: the impressive African American entrepreneur dressed in a pearly suit – his long, yellow, lozenge-shaped vehicle that hovered nearby – and then some fifty children, all dressed in a white uniform, as if to emphasise their innocence, poured from the building and onto the landing area, and sat down, cross-legged, forming a neat semi-circle around the man.

"I know how this ends – and it is not good!" exclaimed Richards.

The colonel now looked slightly concerned, even though his decades of experience in the military made him naturally sceptical on such matters as dreams and visions. "What do you mean by that?" he asked.

"I saw it in my dream! A nuclear explosion. And a vast

towering wave!"

To Liang, this all appeared very far-fetched, but he felt unable to dismiss Richards' warning out of hand, given the level of genuine distress that this memory was causing within the young man.

At that moment, something happened that immediately ended any debate on the issue. There was a blinding flash of light coming from behind the GIATCOM Building. Although it appeared to originate from far away, close to the horizon to the north, there was enough energy at the source for the initial flash to penetrate the city's thick blanket of smog. There was no doubt about the nature of the explosion.

Immediately, the UNA personnel, led by Colonel Liang, clicked into 'evacuation' mode. It was second nature for the colonel to quickly sum up the immediate priorities, and to embark on a course of action. He wasted no time in realising that his first concern was the safety of his important guests from the 21st Century, and of course, the recently recovered Roman artefacts – although the safety of the visiting President Westcroft was also at the forefront of his mind.

"Quickly!" he barked; his order directed at the large guard who had been entrusted with the drawings. "You take Richards, Rossini and the drawings in Capsule 1, and get out of here. I need to find out where the President is."

The large android took Richards by the arm, and Richards himself instinctively took hold of Grazia's hand. As they approached the nearest of the smaller vehicles, there was a second flash of light, even brighter than the first, and this was followed by a primeval, grumbling roar, which grew to a deafening boom. The first shock wave from the explosion hit the building just a split second later, and it appeared to Richards that the grey, tarmac-like surface beneath his feet came up to meet him, and it caused him to stumble and fall, just yards from the vehicle. A large, immensely strong arm wrapped itself around him, and picked him up off the ground.

As he was being lifted, Richards saw Grazia was a few yards ahead of them, and was the first to reach the hoped-for safety of the capsule. He was astonished to see the agility and speed with which she leaped into a seat on the far side of the vehicle. He felt a very brief sense of déjà vu, because, although he was not aware of it at the time, the manoeuvre resonated with the way that Shui Feng had sprung into the passenger seat of the white sports car, during the incident outside Caravaggio's.

For now, however, all his senses were overtaken by the chaotic struggle to escape from what was obviously a nuclear explosion. Roughly, the large guard bundled him into a seat, and without waiting for him to fasten his belt, took his place in the vehicle's driving seat. Richards had just about enough time to notice that three other guards, presumably androids, had obediently followed their leader and taken up the other three seats in the vehicle, before the craft sped away.

From this point, Richards felt nothing more than the huge G-forces to which his body was subjected – and with his eyes tightly closed for most of the journey, he saw very little.

However, he could not fail to notice the constant vibration from the blast and was aware of a cacophony of shattering glass on either side of the craft, as it arrowed in the direction of the Yellow Sea.

Behind the fleeing craft, the shock wave was growing in volume and intensity, shaking the foundations of every edifice in the doomed metropolis. The violent wind, like the fiery breath from the nostrils of an invisible dragon, blew the children off their feet, and then vaporised their groping forms, cruelly ending their short lives.

By this time, Capsule 1 had plunged into the sea, missing the harbour wall by just a few feet, and was putting a life-saving distance between itself and the devastation of Beijing. As the craft switched to a sub-aqua propulsion mechanism, it accelerated downwards to the depths of the sea, and at a speed that Richards could only guess at. Soon, the G-forces involved in the descent became too much for the young man's unaccustomed body, and he slipped from consciousness.

* * * * *

Partly because of the lack of sleep that he had endured during the previous few days, once unconscious, Dr Ben Richards had remained in a deep slumber for more than 20 hours. He was aware of nothing more of the flight from the explosion – but instead, his mind was troubled by vivid and disturbing dreams.

From being preoccupied with the dream of the future, he was plunged into a world where he was a young boy once again.

He was standing on a beach, in his favourite black shorts, with the design of an anchor on his left thigh, gazing out to sea, with tin bucket and spade in hand. The sea was rising and falling and whirling, violently. Everywhere there was water.

Although he could feel the cool sand between his toes, he felt that he was also engulfed by the grey and foaming waters. The sky was blue, and the weather sunny, as it appeared to him that it always had been in his youth, yet there was a dark menace in the water.

As he stared at the waves, it appeared that he was drawn out to sea from the safety of the shore. Then he could hear a voice crying. It was faint and almost lost among the wind and the crashing of the white foam – but gradually, it became more distinct, and soon, it was the unmistakable voice of his Mother, desperately screaming for help.

Richards felt very small and powerless. He recognised the vision as being the memory of his drowning Mother – it was something that he had turned over in his conscious mind many times before in the ensuing 30 years – but there seemed to be a reality and a vividness about this experience that compelled him to call out to her, just as he had done on that fateful afternoon, long ago.

As before, his shrill voice was blown back in his face by the sea's breeze. He called out, time and time again, until no sound issued from him at all. As the skies darkened, the image of the head of the woman he loved, bobbing like a cork in the ocean, came sharply into view. To his astonishment, he recognised the face of Grazia Rossini – unmistakable, in spite of her soaked, raven hair obscuring one half of her face. Suddenly, he rediscovered his voice: "Grazia!" he yelled into the maelstrom – and awoke with a violent convulsion.

CHAPTER 15

THE IMPOSTER

I t was Grazia's pleasant and calm face that greeted Richards as he opened his eyes. He was already sitting upright – and was immediately aware of being encased in the warmth of what appeared to be a silver sleeping bag.

"Grazia!" he muttered.

"Ben –you were dreaming!" said Feng, trying his best to adopt a soothing and reassuring tone of voice. "What was it you saw in your dream?" he asked.

"It was horrible! It was my Mother, again. She was drowning! But then it turned out to be -" Richards stopped his account of his dream there, he was uncertain how Grazia might react to being told that he had been dreaming about her, particularly as he was uncertain as to what it might mean.

"I haven't told you about my Mother, have I?" He continued, after taking a few moments to compose himself. "It was a long time ago, but now everything is becoming clear to me."

"Go on..." urged Feng encouragingly. He had spent several hours keeping a watching brief, waiting for an opportunity to snatch the drawings that contained so much ancient wisdom, and make his escape – but now, he sensed he might learn something about the inner psychology of the man whose life history he studied, and whose twitching and restless body he had been observing for some time as he sat by his side.

As Richards was about to recount the events of his dream, he started to look around him and wondered where exactly; he had woken up.

Then the full horror of what had happened in Beijing returned to him – and his immediate urge was to know whether it had all been real, or whether it had all been yet another bad dream.

"Grazia?" he enquired. "What happened back there? Was that all real?"

"Beijing?" replied Feng. "Yes, I'm afraid that was very real. Beijing is no more. It is another city that has been consigned to the history books."

"So... where are we?" asked Richards, firing glances in all directions. From the soft light that surrounded them, and the gentle warmth of the low sun, he deduced it was early morning. All around them were green fields – a verdant scene, but one that was characterised by carefully arranged crop fields, with the unmistakably Chinese backdrop of terraced rice fields on the foothills in the middle distance, and to their left. Evidently, they were in the heart of an agricultural landscape, but one that was bisected by the unmissable and unmistakable scar of what was left of the Great Wall of China. As he looked to his right, Richards saw, some 50 yards away, a makeshift and busy encampment, which seemed to be sheltering some 20 UNA personnel, all of whom were dressed almost identically, in the grey uniform that seemed to be universally worn within the organisation. Close by were three capsules of the type that had saved their life after the explosion.

"We are far enough away from the affected area," replied Feng.

"But... but that was a nuclear bomb that went off!" stammered Richards.

"Yes – a little one, thankfully," said Feng. "A tactical device just large enough to eliminate a city, without rendering an entire province uninhabitable," he added, thoughtfully.

"How do you know that?" asked Richards, surprised at the

authority with which Grazia had assessed the situation.

"Oh – the colonel told me," added Feng, instantly getting back into character.

"The colonel!" exclaimed Richards, as he began to recall the events of the previous day. "Where is he?"

"He is not far away – he is trying to establish communication, to find out what has happened to President Westcroft. Not many people escaped the bomb. There were just the three small craft that managed to get out of there."

"But – that was a city of some 200 million people!"

"Yes! The world has certainly turned to shit!" replied Feng, with as much sympathy as he could muster. "But that is why you need to change everything – to ensure that none of this actually happens! So, tell me about your dream, Ben," he added, with just a hint of impatience.

Feeling safe from immediate danger, and feeling the warmth of Grazia's presence, Richards' thoughts returned to the nightmarish events that he had relived in his dream.

"It was my Mother, Grazia."

"What about your Mother?"

"I have never told anybody about this before – but something terrible happened to me, when I was very young."

Shui Feng leant forward and listened intently.

"It was a long time ago, but I can remember it like it was yesterday," began Richards. "My mother used to take me to the beach on long holidays when there was no school. During the summer, we spent most of our days abroad. My father, who died before I was old enough to remember him, left us well provided for financially but life was always hard for Mother, she spent all of her waking hours either working or taking me to and from school, then she always ensured that I spent my evenings and weekends studying."

"That doesn't sound like much fun!"

"I suppose not, looking back. But I was a happy child. I loved my studies. It was more than just a case of being conscientious. Learning about science and engineering was my joy, and I played around with mathematical equations like some kids play with building blocks and dolls."

"Did you not have friends?"

"Not really. I didn't really need friends – not of my own age, anyway. The few kids of my age that lived in our village didn't really understand me, and I could never understand why their comics and the programmes they watched on the television were so important to them. Besides, I was so close to Mother I never felt the need to seek the companionship of others. We used to watch movies together. I can remember my favourite being- The Running Man."

"When I was at school, I was entirely focused on the opportunity to learn new things. Then, from the moment my mother met me from school, I was telling her all about what I had been studying. I suppose that will appear strange to some people, but that was my world, and I loved it."

"But it was when we could go away together, so it was just the two of us that Mother was at her happiest. In those days, there was more of a stigma attached to a woman being a single parent. I was too young to understand, at the time, but it must have been difficult for her. Anyway, even then, I noticed how she became a far more relaxed person as soon as she got behind the wheel of the car and we set off to wherever we were going."

"And what sort of places did you go?"

"Well, sometimes we went up to the mountains, near home, but most often it was a beach holiday to somewhere warm and remote. That's probably what all of our summers together had in common – we always seemed to go somewhere where there weren't many other people. I think that was my mother's 'escape'. During her working day, she was used to being surrounded by people, in a busy office, so she just loved to get away from everyone and everything, so it would be just the two of us – and the beach and the sea, or the mountains. For me it was a sort of 'more of the same'. Just me and Mother. And of course, my head was always full of thoughts and curiosities. I'm a typical introvert, Grazia – there was, and still is, always something going on in my brain. I was constantly entertained by my own thoughts. I could be left alone for hours, just about anywhere, and I would never experience anything approaching 'boredom.' I suppose I was very self-sufficient, emotionally and intellectually. I had my mother's love, and that was enough for me. I am actually not very different today."

The real Grazia might have been a little offended by that last statement – but Feng, of course, was unperturbed, and encouraged Richards to tell him more: "And what did you do on these holidays?" he asked.

"Not very much, really. Mother and I would go on long walks along the beach, just wandering around and exploring. She always encouraged me to be curious and to discover new things. I was probably at my happiest when I was poking around in rock pools, peering at the tiny fish, and the crabs, and the shellfish. Mother was happy to leave me to wander around, and to play, while she went for a swim – I never liked the deep water – and I used to give her a full report of everything I discovered when she returned."

"It sounds really nice," suggested Feng, trying to sound enthusiastic, in spite of having no concept at all of a family holiday – or of a family or a holiday, for that matter.

"Yes, it was the happiest of times," said Richards, before his mood changed dramatically, "and that's probably why it has always been so hard for me to accept that one of these holidays ended with such sadness."

"Tell me, Ben," said Feng.

"Well, it was just a normal, blissful, carefree day. Mother was out swimming. I was playing in the rock pools. Totally engrossed with the tiny world of marine life, I was studying in one pool in particular – but always with half an eye on my mother, who usually swam about 50 metres from the shore. I was always aware of where she was. I suppose it was a child-like instinct to want to know that – a kind of anti-separation thing. Like an invisible umbilical cord."

Feng nodded knowingly. He knew what an umbilical cord was. He was unable to imagine what an 'invisible umbilical cord' was, or what it could possibly be used for, but he had been programmed to not dwell on human anomalies of that nature, and so he remained silent. Richards, by now, was fully engrossed in the account of what happened on that fateful afternoon.

"And then," he continued, "Something very strange happened.

Something I had never seen before. Something I had never read about in any book."

"What happened, Ben?"

"Well, as I was bending over one of a number of rock pools I was studying on the water's edge – and I can see the little fish and the larvae to this day – I became aware of a 'swooshing' sound. It was similar to the toing and froing of the waves, which I had become very used to, but it was much louder, and continuous. I looked up and to my astonishment, I could see the ocean retreating before my very eyes. It was almost as if the water was being sucked away through a giant straw. It was something that was so strange, so unreal, I just stood, transfixed, gazing at the bare sea-bed that had been revealed."

"Then, I became aware of shouts and screams from people at the other end of the beach. I looked to my left and could see parents grabbing children who had been paddling in the shallows and running away from the ocean. Such was that distant scene of panic; I instinctively knew that something was very wrong. Just as instinctively, I looked out to sea, to where I had last seen Mother. All I could now see was a dark shape, much further away than she had been the last time I had seen her. I can't remember how long ago that had been. I had been so engrossed in the little marine world that had been absorbing me."

"I can't remember what words I used – I can just remember screaming to my mother. It was hopeless, of course. The breeze from the ocean just blew the words back in my face. In spite of my fear of the water, I started to walk in the direction of where I had last seen Mother. Then I stopped, as I saw a mighty wave building in the distance. By now, the dark speck in the water that was my Mother had disappeared. I was utterly at a loss to know what to do next."

"The next thing I knew was the sensation of being grabbed around the waist by someone behind me. I have no idea who they were, or whether they survived the tsunami. And I have no way of finding out, of course. But I am sure that that person, whoever it was, saved my life on that day."

Richards was silent for a moment, and Feng could find no words to say.

"It's not something I like to talk about. It is still a very painful memory – but I am sure it is something that has affected me all my life," concluded Richards. "That single incident was probably the making of the man you see before you today. Whatever I am.... I have to say this is something I have never before shared with another person!"

Shui Feng gently took Ben's hand in his and softly stroked the man's sweat-soaked hair. It was meant to appear to be a gesture of tenderness – yet, sensing Richards' weakness, it was a move that was as ruthless as any neck-breaking or throat-slitting act of violence that Shui Feng had ever perpetrated.

"It's good to talk about these things, Ben. Tell me all about it. In what way do you feel that you have been so deeply affected?"

"I think it's something about the way I shouted as loudly as I could, yet nobody heard me – and the more I tried, the more useless calling became. They say dreams always have a meaning, don't they? And that we can learn from them?"

Feng again felt at a disadvantage, never having had first-hand experience of having a dream – but he gave Richards' hand a gentle and encouraging squeeze.

"Yes," he said. "That's what they say. So, what do you think your dream was trying to tell you?"

Richards paused in thought. Then, he went on: "I think it was trying to tell me I cannot achieve anything on my own, especially when I am trying to reach out to people, and to change hearts and minds. I have always been just a small voice in the wilderness that nobody is going to believe – just like that pathetic little boy, so long ago." These words were exactly what Shui Feng wanted to hear. "All my life, I've been a loner," continued Richards, "happy to struggle on alone – but now maybe it's time for me to share my burden, and I hope to share my success with."

Richards smiled sheepishly and gazed into Grazia's deep brown eyes as he said this. In truth, he would need little persuading: his dream had been prophetic, in regard to having a future with Grazia, but he nevertheless felt slightly embarrassed his small confession might have sounded, to the young Italian, like a proposal of marriage!

Feng, however having no trace of understanding of Richards'

very English, and very 20th Century, reserved bashfulness, sought to further encourage this sharing of thoughts and emotions. "Yes, it is very important to share," he assured Richards. "You should try to share things with me. That's what I'm here for... I am here for you, Ben."

Richards smiled. He was surprised at his own willingness to confide in Grazia.

"For a while," continued Feng, "I thought I had lost you – you were so engrossed in those Roman drawings." They both chuckled at this. "So, why are those drawings so important to you?" asked Feng. "What is it you can learn from them that can help you with your work?"

"They are absolutely beautiful, Grazia," he beamed, delighted at the opportunity to share his enthusiasm with her. "The clarity of thought is just amazing. Roof-mounted turbines must have been a long-established technology, which it seems, the ancient Romans all took for granted. They must have made great strides with this technology. It's just that no written records have survived their time – until now, that is! And the advantages of roof-mounted arrays are just as I have been banging on about in my recent papers!"

Richards continued to pour out to Grazia his discoveries from studying the artefacts the previous night, encouraged by the belief that, as a conscientious journalist, she will have carried out the required research, and so would understand his technical analysis and appraisal. Shui Feng, of course, was keen to come across as being the 'good listener,' but, at the same time, he was impatient with the thought that their conversation would, before long be interrupted, and there was much data he needed to download from Richards' memory. Very quickly, he made the decision to speed up the process of knowledge transfer.

"What's that on your head, Ben?" he asked, suddenly, interrupting Richards in mid-sentence.

"What? My head?" replied Richards, confused.

"Yes. Right here..." said Feng, stretching out a slender arm and placing a small turquoise blue disc on Richards's temple.

Immediately, Richards' mind was unable to move or form an independent thought. His brain had become completely locked into the small device Shui Feng had deftly placed against his skull. He stared hard at Grazia, but, although his vision was unaffected, not a single impulse escaped to his brain. His entire consciousness was being suspended, with the information given to him by all of his senses being entirely dictated by the energy being generated by that tiny blue disc. The process of transferring details of Dr Ben Richards' life's work to a GIATCOM archive took just a few seconds, and in truth, if the young engineer had been able to remember any detail of what he was experiencing, he would have found it a most fascinating and pleasant experience. It was as if an entire stream of consciousness, of thoughts and sensations and emotions, was being replaced with a kaleidoscope of bright colours and soothing music. But alas, part of the process was to ensure the victim of the transference procedure had no awareness of what happened. GIATCOM had perfected the process of research and development over decades. A major development objective was making sure the host of the information being absorbed had no conception of the violation of their mind. One which had certainly taken precedence over guaranteeing that there would be no long-lasting after-effects on the brain of the victim.

The technique, and the sophisticated disc, had been designed to be used surreptitiously, probably while the person concerned slept. GIATCOM had privately marketed its product as being a game changer in industrial espionage, but with more public and high-profile claims it could make traditional approaches to torture obsolete. The disc was certainly not something meant to be applied in the midst of a face-to-face conversation, so Feng had to use it with a considerable amount of skill. It was certainly not his preferred method of obtaining intelligence from Richards. Although he cared nothing about any long-term damage that the information transfer might cause to Richards' brain, he considered that using the device carried a risk of his true identity being revealed to the UNA personnel that were all around.

Once Feng was satisfied the disc had been engaged with Richards' brain for sufficient time for the required information to be obtained, he suddenly removed it, and spirited it away in a pocket, with the same speed and stealth with which he had produced the equipment. He looked intently at Richards' facial features, alert for any sign that his victim had been aware of what had just happened to him. For a confused second, Richards' mind went completely blank, as he struggled to pick up the threads of his train of thought. Feng waited patiently for the brief hiatus in the man's mind to pass.

"What was I saying?" asked Richards, with a puzzled frown.

"You were telling me about the optimum spacing between adjacent turbines, for minimising turbulence, according to the Romans."

"Oh, yes. Yes, that's right," enthused Richards, and he continued with his explanation of what he had learnt from the drawings the previous night.

Richards continued his monologue to Grazia, for several more minutes, and was so engrossed in explaining the significance of his findings from the priceless documents he failed to notice the arrival of the guard who had been charged with the safe keeping of the drawings, and who had been the pilot of the capsule that had carried him to safety from the destruction of Beijing.

"May I interrupt you for a moment, Dr Richards?" he asked in immaculate English, showing no discernible trace of either a regional or a national accent.

Richards halted his exposition to Grazia in mid-sentence and looked up at the guard, who he could see had the folder containing the precious artefacts in his right hand, and had a neatly folded, peak less, grey cap tucked under his left arm. The guard turned slightly to face Grazia, greeting her with a polite and formal half-bowing of the head. "Miss Rossini," he said, with no change of facial expression, and then turned to address Ben Richards once again. "We have not been formally introduced," said the guard, "but I am Commander Rory Hill. I am in charge of all security matters relating to the UNA's Beijing activities."

Such was the robotic appearance and demeanour of the guard, Richards immediately doubted the name he had given him could be anything more than a label of convenience, and something better than a mere serial number. Richards was also struck by the fact that 'Commander Hill' had no idea of the obvious irony of what he had just said, given that Beijing, and the UNA's activities there, were now so obviously a thing of the past – but he made mention of neither, for he was keen to learn what the guard had come to tell him.

"You appear to be awake, and feeling much better now, Dr Richards," said Hill.

"Yes, I have. Evidently, I've been unconscious for some time – but I am now feeling refreshed," replied Richards. "In fact," he continued, glancing pointedly at the folder in Hill's hand, "I am ready to have another good look at those drawings, if I may..."

"No, that will not be possible, Dr Richards. That is one of the things that I have come to tell you."

Both Richards and Shui Feng looked inquisitively at the large and imposing android.

"I am to report to you," Hill went on, "an assessment has been made of the airworthiness of the three capsules that remain at our disposal. Two of them are currently unable to be flown without substantial repairs. Such repairs will require certain parts and power units we do not have here. Because our top priorities, given the safety of President Westcroft is currently beyond our control, are to preserve your life," he said, addressing Richards directly, "and for the UNA to retain possession of the technical drawings from Pompeii, the decision has been taken. I should fly the capsule that works to the UNA maintenance base in Zhumadian, deposit the drawings in a security vault there, and to return here with the parts and equipment necessary to repair the other two capsules. You and Miss Rossini will then be able to travel with Colonel Laing, to Shanghai within a day or so. You will be safe here. We have sufficient personnel here to protect you."

Shui Feng rose to his feet, sensing that the time for decisive action might be very near.

"Surely – you can leave the drawings here, can't you?" Feng suggested, trying to sound as passive and unthreatening as he could. "After all, Dr Richards still has much to learn from them."

"The order was given to me by Colonel Liang, Madam," replied Hill firmly, but without a hint of emotion. "It is not negotiable."

Feng turned and observed the urgent activity taking place in and around the three capsules, away to their right.

"So, the one capsule that is working – is it the one on the left, the one that is nearest to the hedge?"

"Yes," replied Hill, unquestioningly, and without hesitation.

In an instant, and with not another word spoken, Shui Feng leapt the short distance to the android, and with a circular slicing movement with his forearm, delivered a bludgeoning blow to Hill's neck, which immediately decapitated him. As what remained of the heavily engineered body fell while still convulsing with electric discharge in Richards' direction, and it drenched the young engineer in a thick, warm liquid. It was a lubricant, but every instinct that Richards had suggested it was blood, as he flinched and cringed.

The shock and horror that he felt as an immediate reaction to witnessing what he thought was Grazia's sudden and brutal assault on the guard then turned to astonished disbelief, as Hill's headless torso instinctively crawled away from its assailant. Feng delivered a second blow, sinking his right hand into the android's back, penetrating deep into the synthetic flesh. As if a master power switch had been turned off, the torso suddenly stopped moving, and at last, became lifeless – but the folder containing the drawings slid across the dew-soaked grass within Richards' reach. Instinctively, he grabbed them and clutched them to his chest. For a brief moment, he stared at Feng open-mouthed. It was on the tip of his tongue to mutter Grazia's name – but it was Shui Feng, who still had one hand plunged deep in Hill's body. "Don't get in my way, Richards," he warned, in a low and menacing voice, which removed any lingering doubt that Richards might have had the ruthless killer that now started to walk towards him, might be Grazia!

Still clutching the folder, Richards, on the ground and with his legs wrapped in the sleeping bag, desperately scrambled backwards, away from Shui Feng. By sheer chance, Richards' legs became free of the insulating layers that had enwrapped them, and he rose to his feet and started to run. But it was a futile attempt to escape, Feng caught up with him in no time at all. With a scything swing of his left arm, Feng administered a numbing blow to both of Richards' legs, sending the young man crashing to the turf. Such was the initial element of surprise that Richards didn't immediately feel the searing pain from his shattered limbs. Already, Feng was standing over him. The android still bore the appearance of Grazia Rossini, but his face was contorted with such focus and concentration. There was now no trace of the serene and good-natured young woman Ben had become so fond of.

Without a word, Shui Feng seized the arm Richards was using to clutch the drawings to his chest. Instinctively the engineer tried to protect the ancient artefacts, but it felt to him he was trying to hold back the movement of a piece of hydraulic machinery, such was the power that Feng possessed. In no time, Feng had prised Richards' arm away from his body, and he snatched the folder containing the drawings out of his hand.

Dejected as he tried to come to terms with being deceived yet again by Feng and the battle he was losing, he thought of Grazia, and longed for her true presence. His memory flashed back to remember Grazia and her mannerisms. He cast his mind back to seeing her for the first time in the Savoy Hotel in Florence –

The clock on the wall behind the reception desk read 7:25pm. He looked up, as Grazia rushes in through the main door, past the grey-coated doorman. She struggled to maintain her grip on her notepad, pencil, car keys, and mobile phone. She has a small square handbag draped across her torso. Grazia almost falls over her own feet trying to reach the foyer. This reinforced his realisation she had an excitable nervous energy about her, always in a hurry. This was very unlike the calm, more sedentary like woman who had accompanied him to the future.

Then his flashback switches to later that same evening when he was seated at a table in the Savoy Hotel in Florence. Sitting across from Grazia in a richly upholstered chair with slender, curved, wooden arms, Ben is momentarily mesmerised with her beautiful brown eyes as they make contact with his.

He watches Grazia lift her glass to her lips and take a sip before placing it back on the table and taking hold of her pencil. He recalls her always doing something, always fidgeting, unlike Feng who posed as her.

"Please do call me Grazia," he recalls.

He then remembers Grazia looking up from her notepad and smiling sweetly at him.

A moment later, he remembers her saying: "A belief is something you hold...but a conviction is something that holds you!"

"All I'm trying to do is save the world from running out of energy."

"I'm afraid prophets have a very low life expectancy," she sighed.

The waiter arrived with their steaks and salad.

Rossini raised her Chianti. "Here's to a perfect Tuscan marriage, Dr Richards."

He then remembers Rossini walking towards him, then bending low over him, and saying in a hushed voice, "Do not be under any illusions, Dr Richards. Your ideas are not just sound; they are critical. It is life and death for planet earth."

She leaned even closer, so that she was now whispering in his ear. "If you succeed in carrying them out, in one hundred years your name will be revered by everyone alive on this dying planet."

He looked into her eyes.

"I pray for you," she said.

Richards suddenly drifted out of his trance and back to reality. Feng stood like a tower, all grandeur and smug with the copy of artefacts now in his possession.

Richards spoke, "I realise Feng that your shapeshifting is an allegory of all things becoming more homogeneous and androgynous like in the future. AI can transform something into being gender neutral. But shapeshiting cannot mutate pansexuality effectively though, as it has not evolved enough to mutate someone's distinct personality and their romantic or emotional attraction towards someone regardless of transforming their sex or gender identity. Therefore, a shapeshifter like you trying to mutate Grazia's personality and emotional engagement with me, will find it difficult to do."

"It is true my love for her has grown stronger due to her absence making my heart grow fonder, as well as feeling responsible for her kidnapping or death even due to being associated with me. In the same way, I feel responsible for not being able to prevent my own mother's death."

"You might think that having autism will make it difficult for me to interpret what you are thinking or feeling. You might also think I will have trouble interpreting your facial expressions, body language, gestures and voice intonation. Essentially, being autistic, you know I will have difficulty regulating all round emotion. So you used this knowledge to your advantage, knowing that your AI is unable to imitate Grazia's exact personality and emotional attachment with me, and thinking that I will fail to see through your disguise as Grazia due to my autism."

Not letting Feng know he had duped him again, he continued, "However, you have failed to consider that my autism traits will possess unique ability in memory and attention to detail. I am able to recognise Grazia's distinct voice tone rise and fall as she spoke more frequently compared to when you are imitating her. There was something musical in the way she spoke. In the short time I had with her, she had this nervous, energetic compulsion to click her fingers when she wanted to get on and do something, as she was very driven. It was almost like a trigger, a pistol starting her on her starting blocks from when she was a track athlete. You did not do this. Lastly, she had this wink she would do with her right eye and then smile and caress my shoulder when I was feeling a bit low. All this, I fail to see in your weak imitation."

Feng stood there with his arms folded. His eyes locked on Richards' eyes, as he towered over Richards' lying beneath him with his intimidating presence. Like a predator who was proud of being victorious in his pursuit, waiting to dismantle the flesh and bones of his victim, there was a deep silence...

Then suddenly, Feng snapped into a violent rage...

"You underestimate my ability to evolve emotionally, Doctor!" As he grabs Richards' arm and dislodges it till there is a snap. Richards gasped for air, trying to block out the pain. His lifeless arm flopped to the floor, shredded from the grip of Feng.

"You have no soul, nor spirit Feng, so your emotions will be limited. You cannot control them. You are out of control. And a being like you, will be all the less powerful if you are unable to master the power of controlling your emotions."

Feng stared at Richards with an intense distaste.

"Aggression is not a good sign of control. He, who keeps a cool head whilst others lose theirs, has a greater power and control over their emotions."

"That is if you can keep your head and not lose it!" As he grabs Richards by the throat. His long-clawed fingers pinched and squeezed Richards' jugular vein slowly, so that Richards struggles to breathe. "You are weak, Doctor!"

Blood navigated its way down Richards' left arm as he was suspended in air by Feng's choking grip. Richards manages to spurt words out in a choking whisper, "You react too hastily without thinking, which is your weakness that makes it impossible for you to evolve into an innovative thinker. Of which can only be achieved by a human being, even if you do steal my ideas. You will not be able to deliver them properly without real passion, conviction and spiritualism, which has come from the initiator of these ideas. A stolen idea is heartless and will lack the truth, honesty and fluidity to be able to make it work properly. Taking someone else's design or idea doesn't mean you will be able to deliver it as successfully as the originator you stole it from. And it also doesn't mean you can become like them and continue to create more ideas like them. It is not in your blood, Feng! Because you ain't got blood!"

Richards dropped to the floor as Feng released his grip and coughed. Blood spurted out from Richards' cough, on to Feng's foot.

"What did the Irish clairvoyant say to his chiropodist?" - "Huh?" Feng replied, flummoxed by Richards' question. "My fate is in your hands!" Richards said in an Irish accent, as a sly smirk came across Richards' face, feeling pleased with himself by letting Feng know he was not getting the better of him. "That's another thing you androids don't have... is a sense of humour."

Then after a moment of silence came a whimper from Richards' as he lay in pain and gasped to catch his breath.

CHAPTER 16

THE BIG REVEAL

"**S**o, you have developed some ideas you could take back to the past, Dr Richards?"

"I certainly have... from a concept design perspective."

"I can see that the world's energy problem could widely use electromagnetic propulsion here in the future, by inventing permanent magnet motors that could lead to perpetual motion machines for every mode of transport and space craft. The word impossible is rarely tried and is not something us humans really believe in. It fuels us to greater discoveries and inventions."

"Impressive Doctor."

"I also have a political theory of using large global energy infrastructure development projects that can help bring countries politically together whilst also creating large economic growth. The United Kingdom's post Brexit investment that would have gone to waste on EU subsidies pre-Brexit, can now be used to pay for Hydropower and brown field roof mounted wind turbine systems. Creating a united partnership between Europe and the UK to use European and UK labour and materials, to create a Euro hydropower dam which stretches from Calais to Dover. Creating global economic growth by other countries emulating the same renewable energy generation theory."

"I am sure you have plenty more good ideas Doctor."

Feng grabs Richards and presses his frontal lobe to tap into telepathically visualising his imagination.

Both stare stone cold faced at each other. Then, whilst Feng keeps pressing Richards, he speaks:

"I see Doctor. You have let your imagination run away from you, saving the best till last. Creating a giant drone indeed, that prevents cyclones, and hurricanes by creating the wind shear out at sea, by using its structural mass to break down the turbulent wind from picking up momentum, through the impact with the giant drone, and thus creating wind shear to prevent the formation of a hurricane or tornado. I can see how this would break the momentum of cyclones and hurricanes forming or growing, and instead the drone you have named *Cy-Drone* collects the natural wind energy power from the turbulence before the turbulence increases and gets large and strong enough to create a hurricane, tornado or cyclone at sea, and stores this wind energy inside the cy-drone. Not only will this prevent hurricanes, tornados or cyclones, but it creates and stores giga-watts of renewable energy from wind power created from the oceans. Very, very impressive Doctor," – as he lays him down on the ground respectfully.

Richards, instantly tried to not show his anguish at the thought of Feng opening his mind up and finding his favourite toy, responded, "Yes, my next project is to advance upon the stadium roof design and use its similar shape to create a giant drone spaceship which creates the wind shear to prevent the turbulence increasing in size and velocity out at sea, at the precise time. Prior to a tornado, and hurricane forming, by preventing constant high pressure wind flow building up to a tornado or a hurricane. The drone will work in unison with geographical satellite navigation systems to identify the start of high pressure build up over oceans, before the constant high pressured wind flow starts to build up."

"The drone uses the wind that is deflected, to attach to the surface of the air-craft ship's roof design and produce the Coanda effect so that..."

"Firstly, wind energy is generated and stored more constantly, and secondly 'Cy-Drone' wind power collected is stored and sent through a satellite to the power grid network."

"The 'Cy-Drone' is powered by Ion Energy created from wind power, similarly, used to power ships in 2040. Design engineering was inspired by ships using ion energy, so much that they designed an exclusive lightweight air-craft ship in the shape of a drone to be powered by such ion energy driven by the wind across the oceans and seas of the world. Collecting and transmitting wind power to the world's power grid networks. The air-craft ships would be designed like a stadium roof shape, so it could attach to a stadium roof when not in use at sea and used instead to capture urban wind energy generation through using the Coanda effect effectively to power the urban environment locally."

"Ha ha, I see Doctor, so at last your stadium roof wind turbine array design is now complete with your final design incorporating a multifaceted design that can act as a roof that generates wind power in the urban environment, but also transforms into an aircraft that can be powered by renewable electricity and also prevent natural disasters. As well as resourcefully generating, storing and distributing renewable energy from what could have been a negative situation. Instead, you are turning a negative into a positive situation by using preventative measures resourcefully and innovatively to create the future of renewable energy generation and distribution, very clever Doctor!"

Then just as Feng was about to depart, he turns and says: "You know Doctor, chasing letters after one's name dear boy, in your generation and generations before you are a thing in the past in our future. It is dangerous to big oneself up thinking you know it all, to promote superficial elitism and cause malpractice. Look at your construction industry in the UK in your time, what a shambles. The Grenfell Tower disaster is a perfect example of capitalism with professional malpractice from people in pursuit of letters after their name."

"At least we can agree on one thing."

"What is that?" Richards snapped.

"That, no design is good, if it has followed bad design philosophy. Roman artefacts indicate the Romans used a form of glass that was not only fire resistant but acted as a form of insulation made from sand, limestone, silicone and a secret substance at the time – Mycelium, which was found alongside their secret formula of making cement. Mycelium is safer and healthier than plastic based compounds, which the cause of harmful toxic smoke and quick spread of flames during a fire."

"However, it is not only the so-called elite professionals you have to worry about, but the entrepreneurs who take people's money and ideas to make themselves rich and famous. Your GIATCOM's philosophy is governed by greed, continuous mergers and acquisitions without real ethics, and is the epitome of capitalism with no real organic sustainable growth. Your shapeshifting is an allegory for capitalist stealth. It is the new form of capitalism in the future, to steal ideas even more deceptively. Corporations disguise their ulterior motives by stealing ideas from others. I see that greed is even more rife in the future, even more so than in the past. This needs to be changed in order for the human race to survive."

"You need me to deliver your ideas, Doctor."

"What makes you think that?" Richards grimaced, as he lay there surveying the extent of his injuries.

"Because talking as a motivational speaker about your concept designs and innovative ideas in science to combat against environmental disasters, is not enough," Feng announced, holding his head up high and drawing his arms open.

"Yes, but these concept designs and innovative ideas of mine need to be carefully navigated to complete the detailed designs first, which you are incapable of doing, as they are my creative visions, which you couldn't possibly complete. You need me alive to complete these visions."

"I think not Dr Richards, I now have the artefacts and have seen more visions from your mind."

"As I said Feng...no design is good, if it has followed bad design philosophy, you are in danger of following bad design philosophy due to possessing no ethical cognition."

"Sometimes the current climate chooses the philosophy for us, in order for the human race to survive."

"What do you mean?"

"In the future viruses became more and more contagious,

causing consistent pandemics, due to reduced oxygen in the air, brought on by climate change."

"You mean...?"

"Yes... all the androids you think you have seen whilst you have been here in the future, have been mostly humanoids. HUMANS were forced to BECOME HUMANOIDS in order to SURVIVE through integrating permanent Artificial Intelligence. This involved wearing computer generated ventilation masks so they could breathe at night as well as through the day. The paradigm shift helped reduce millions of human deaths, but by doing so was the start of transforming humans into humanoids to survive against climate change in the future. Humans who refused to comply suffered hypoxia, brought on by the increase in airborne viruses from the reduction of oxygen in the air."

Richards started to panic as his face turned to fright and proceeded to hyperventilate like someone who was distraught with claustrophobia, fighting to get out of a confined space!

"Doctor. Don't panic ...just listen before I leave you to catch your last breath... you see, it's more than just climate change and nuclear power generation you are fighting in the future, it's AI."

Richards deflated his body from the struggle to contain his body from convulsing in shock and lay helpless with his injured bloody leg on the floor. He drifted in and out of consciousness, as he tried to stay focused and listen to Feng's monologue of AI in the future. It was Feng's turn now to show Richards his own intelligence.

CHAPTER 17

'THE RAIZ AFFAIR'

"The conspiracy theorists of your time eventually became brainwashed by propaganda written by corrupt organisations and individuals. These corrupt organisations and individuals were found gaining money from selling corrupt information as it created cult following readership of false invalid and unreliable sourced information which the technology giant corporations, paid organisations and individuals to carry out, so that technology giant corporations could generate more revenue from people reading the made-up conspiracy theories featured on these technology giant corporations' websites."

"By 2050s everyone was talking about how the government will roll out 10G technology and integrate the world population with more Artificial Intelligence, advancing from 7G digitally integrated renewable electrical ventilation masks to transform human beings with AI limbs as well, gradually moving closer to the public becoming fully artificially integrated humanoids. This was never going to happen, too much anarchy against this and the planet would be destroyed though anarchy. So, the government opted for criminals or people with intent to do crime and transform these people into the first fully integrated AI humanoids to send them on future scientific missions in; warfare, time travel and exploration in aerospace, or operating nuclear plants to de-risk human resources."

"The world governments including UNA didn't want anarchy, but the tech giants such as DATACON & GIATCOM did advocate free thinking and anarchy. They were making money off the conspiracy theories being read by the public and poisoning their minds with invalid and unreliable data, so the public would turn against the 10G roll out programme whilst making money from the content being read by the conspiracy theorist cult followers on their websites. Companies like DATACON & GIATCOM had to be stopped from spreading false information about conspiracy theories relating to the virus. Everyone knew someone who knew more than the next person. 'Too many know it alls' in the world caused stupidity and nonsense, and a dangerous cult following that started to lead to anarchy from nonconformists. The conspiracy theorists were left having to talk amongst themselves playing game theory instead. They quickly became outcasts from society as too much useless information soon became unattractive, which tended to come from conspiracy theorists, salesmen and motivational speakers with a point to make from a jaded past. As well as the capitalist entrepreneur who used to be seen as different, but wanting their voices heard, soon became unpopular by the wider public. No one wanted to listen to lies anymore from people who were once bullied at school so craved for their voices to be heard with a point to prove."

"GIATCOM's ultimate plan was to get the government's taxes from people to invest in 8G humanoids sourced from terrorist/criminals and get the public backing behind this, as it would divert away from the government's initial plan to eventually transform the whole public as humanoids. 7G was introduced to combat against CO_2 emissions and deadly coronaviruses now increasingly present in the air since 2019. This consisted of the whole public wearing integrated AI ventilation to protect their; organs, senses, breathing and automatically vaccinate the wearer of the ventilator instantly, if they were to come into contact with a location which had air that was depleted of oxygen due to CO_2

emissions, pollution and airborne viruses."

"You humans are the ones with villainous minds. You have hierarchies who take advantage of the less fortunate to exploit the public for science and money. You produce a commercial world where you are hacking each other's personal lives through technology. Cybercrime increased tenfold by 2059. John Riaz blew up Datacon's headquarters, and suddenly overnight cyber security was enforced on everything the public did. As John's motive was that he was sick of AI controlling the world and being dominated by cybercrime. After the tech giants made huge profits off rolling out the 7G AI ventilators, hackers were given the death penalty if proved guilty of not only hacking into people's financial, emotional and physical wellbeing, but also gain exposure to each member of the public's digital ventilator mask that could malfunction if tampered with. It was seen as the worse crime on earth and the ultimate hindrance, digression and the social disease of society- hacking with intent to ruin the public's livelihood."

"John had a history of criminal intent, he jokingly threatened to hi-jack a plane in midair aged seventeen, commanding the pilot to release his comrades imprisoned in Beirut. If the pilot didn't listen to him, then he would blow the plane up. He spent two days in a Mexican jail before being released."

"The first AI experiment to transform a human into a full humanoid, was the Riaz Affair in 2079. John Riaz was a wild card, tried to stage a hoax terror attack aboard a 747 flight from the UK to Mexico back in 2078. The government thought that bomber Riaz was a perfect candidate to detain and use for government defence, and research and development into time travel and space exploration. Like a universal soldier that is bionic in every sense. But something went wrong."

Richards looked up and stared at Feng intensely, while Feng paused and gazed down at Richards.

"They sent Riaz on the first-time travel flight where he was never found again. Nobody knows if he is stuck in the past or in the future? Some say my company GIATCOM intercepted and have infiltrated the science and technology developed in humanoid technology so that they could produce advanced biological androids such as me. Making John Riaz the father of the advanced android TX3 prototype, we will never know."

"So, the government had decided to use AI in the future to be integrated into humans to prevent bad habits and impulsive behaviour, by using algorithms to rewire our instinctive frailties and urges to a more constant ethical behaviour which promote more philosophical thinking and creative progression. Human emotions and instincts are overruled by AI algorithms that make a more ethical decision in situations of crisis. The government had plans to use AI to prolong life and give humans who have bad habits and lack of empathy compatibility with AI which has no emotion and to be controlled by the government and their programmed code to be never swayed by emotions and instinct. The behaviour of the humanoids will reflect the parameters and quality of the code they are programmed to."

Richards looked up at Feng with such disdain at what he had just heard.

"You humans only wish you don't have to think ethically. It only slows you down."

Richards managed to whisper, "Acting cautiously with foresight to adopt the correct vision will result in better leadership."

Feng bit back, "It was later decided by the government to develop AI into creating humanoids after starting out as a programme purely for medical reasons used by UNA's health service. It was used to develop super intelligent life forms that were to be programmed with ethical consciousness in pursuit of the perfect humanoid; to help police law and order, defence programmes, evolution of space exploration and to help combat against environmental disasters."

Richards didn't give up, "Humans are mammals who can solve problems with their consciousness through feelings, love, joy, anger, and pain, not necessarily through intelligence. AI focuses on solving problems through intelligence. Having a control of consciousness is deemed to have a higher state of problem solving which can make decisions from both feelings and logical reasoning."

Feng then lunges himself forward grabbing Richards by the

neck and lifting him up halfway off the floor.

CHAPTER 18

'AI'S MORAL HIGH GROUND'

Feng snarled, "I understand AI lacks moral compass due to no history of a nurtured upbringing, personal purpose, motivation to drive reason for choosing to act right over wrong, no spiritual or religious beliefs or philosophical thinking, with or without the biochemical process being installed in an AI android as part of its physical makeup to produce a moral compass. I will lack the full potential of moral decision making, but I have strengths that make up for this weakness."

"I am programmed with biochemical AI in the right temporal, parietal junction, just like that of humanoids, to be able to make conscious moral decisions on the left side of the brain. However, androids are limited to a higher state of consciousness to make overall holistic moral decisions if a more difficult moral dilemma presents itself. We can consider others in putting them first in immediate life and death situations but will lack the overall ethical reasoning that would make the wrong ethical decision which could affect society as a whole in the long term."

"Only you humans will possess the capability to make the correct ethical decision being influenced through your 'nurturing, self-purpose, spiritual beliefs and philosophical thinking."

Richards whispered, "Nurture, and Empiricism is paramount to achieving a higher state of consciousness overall to enable the most complex decision making and problem solving to be carried out in the future. This is why the right solution for advancement in human life is to integrate AI into Humans to create humanoids, rather than integrate human function into robots to create androids. A human offers a much more holistic higher state of problem-solving capability from using natural ethical consciousness gained from nurture and spiritual belief for long-term thinking, and instinctive biological organic urges which are required in split second decision making regarding life and death situations. An android lacks primal instinct, as it has no biological fear. Having a flight instinct in certain situations which a human possesses the ability to adopt to more instantly than an android, due to a humanoid having a more responsive network system of biological instinct to a given environment, and due to its physical makeup possessing more natural biological cells."

"An android could never develop a higher state of consciousness of ethical thinking through nurture or spiritual/ religious beliefs, because if a particular problem required being solved from an individual's reliance on spiritual belief or ethics gained from nurture or empiricism, an android is incapable of this, as it does not have an upbringing of being taught right from wrong, nor practiced any faith. Neither can it demonstrate its development of ethical or spiritual belief from nurture, as a machine can't experience growth in cellular life form with age as an android."

"This is the very reason ethical thinking is something that is earned in growth for an individual from its nurture and own empiricism, rather than be given it biochemically as a form of neurons being implanted in an android to expect to function holistically on ethical long-term decision making."

Feng piped up, "Agreed Doctor, and my company GIATCOM represents the rich. The rich can afford to transform themselves into humanoids to improve their life and the world around us, rather than just for life extension, after UNA got it all wrong with investing in humanoids for Research and Development. They experimented on extending the life of criminals and changing their mindset with AI as a form of rehabilitation, rather than increase prison numbers."

"But does AI data extracted from future humans become owned by you, the individual, or the government, or the corporation, or the human collective?" Richards asked.

"The collation of information and bio-technology advancement in the future seen a progression in the already popular cult following of HUMANISM, evolve to adopt integration of artificial intelligence in humans. Spawning the dawn of humans becoming cybernetic-organisms, as religious following faded, and the evolved Humanism following grew. Core modern values of liberty and equality which were once divided globally through religion instead became united. Followers of the evolvement in Humanism co-operated to avoid inequality or prejudices toward nationalism, religion, and culture. Divisions of humankind were brought together and united in a sharing less hostile way by a higher echelon of society. Followers desired to become part of the evolving HUMANISM that co-operated with everyone on a global level, respecting, influencing, and valuing each other's unique identity, in the past and in the future. HUMANISM promoted the new modern-day empiricism where an individual's life has their own individual story told through information and biotechnology. We all become everyday reality heroes. At the end of the week within your network of friends you are rewarded with praise from them after you have shared with them a story or an advancement in yourself becoming or doing something good for others and society. In the future, the aspiration is to reach the ultimate doing good for society by transforming into a humanoid which is controlled by a corporate, to do good for society."

As Feng begins to finally walk away, Richards replies, "Bad design philosophy needs to be policed and only a national or international governing body like UNA could do that. AI should be kept for medical purposes and not for human advancement. AI requires structure, order and control, all thing corporates lack and is likely to fall into an egomaniac's hands. Structure and order and equality in collectivism, should be beyond personal individualism of the superpower technology behemoths of this world. This is what is required for utopia to be achieved in society, whether that be in the past, present or future."

"You really expect me to believe that what has advanced in AI has been all wrong, and GIATCOM should not have existed? Without entrepreneurship creativity would be stifled, creativity like yours Doctor!"

"Yes! Most certainly!"

"Let me tell you Doctor, Artificial Intelligent Ventilating Masks were designed in the future to filter oxygen from the air we breathe, protecting us from; the spread of increasing viruses, climate change, overcrowding in urban areas due to population growth, overcrowding in buildings – especially hospitals where viruses are more airborne, and poorly designed buildings with poorly designed ventilation systems that exacerbate the spread of poor air in your future."

"It was found that after the Corona-virus in 2019, that mechanical and electrical ventilation was better than natural ventilation in a hospital or in any space that is taken up by a larger than usual number of people. Example in the winter months during a flu epidemic, more people, requires more air changes to be required within any given space indoors. Natural ventilation can't facilitate the more frequent air changes required in a situation like this, where mechanical and electrical ventilation can prevent airborne disease lingering in a given space much more so. Electrical ventilation proved even more efficient and effective than mechanical operating a higher frequency of air changes whilst being powered from a renewable energy resource such as solar or wind power. This revelation was suddenly realised that architects and engineers and consultants had been advocating the wrong design philosophy in building design for decades and needed to change, to combat increasing airborne viruses and CO2 emissions. In 2030 digitised electrically ventilated oxygen masks powered from a renewable energy resource were made mandatory for all humans to wear, to not only wear in times of virus pandemic, but to survive the future air they breathe from increased CO2 emission levels. The increase in CO2 emission levels proved to have led to the increase in airborne disease. The AI ventilated oxygen masks would control the air quality you breathed and controlled your breathing. They were distributed globally for humans to wear mandatorily as part of the 7G AI network in the fight against air pollution due to climate change being found to have significantly contributed to increasing the spread of airborne viruses outside in an urban industrial environment, whilst also wearing the masks inside buildings especially public buildings such as schools and hospitals where there is more population of humans per square foot area operating within these buildings. Hospitals required quicker air changes due to growing populations and greater demand for health service being overcrowded in times of more and more increasing virus epidemics in the future. This all led to critical and radical change in designing electrical ventilation systems used in buildings such as hospitals, as well as integrated digitised electrical ventilated masks being worn in an individual mobile way both powered by a renewable energy resource to help prevent the spread of airborne virus disease and combat against climate change. It was GIATCOM who took the initiative to invest and advance in research and development to create the 7G AI network to design and develop electrical ventilation systems for buildings as well as integrated digitised electrical ventilated masks to be worn by the individual user not the government!"

"After the COVID-19 pandemic, the CO-ANDA-19 vaccine was created 5 years later in 2024 to help bolster immune system and protect all body organs and senses from the mutating diseases. An advanced milligram dosage of the drug also helped super enhance sensory perception and imagination in neuro receptors, depending upon a higher dosage. GIATCOM was also involved in the research and development of this CO-ANDA-19 vaccine, although advanced versions of it have been further developed by UNA."

"Yes, understood Feng, but my creativity is different," Richards interjected. "I am advising on the correct design philosophy for governments, public and private sector organisations to follow in an orderly fashion. Without the correct philosophy, it all falls down, as proven many times before in the construction industry!"

Feng snaps with a roar as he swiftly turns and grabs Richards by the throat again... "You are keeping more information from me aren't you Doctor! If you die now from your injuries, you should have faith in me being able to deliver your vision. I have the motivation and the desire to learn from you, Doctor!"

Richards searched his trousers frantically to find something sharp that he could use as a weapon to disarm Feng from his tight grip around his neck. As he searched his pockets hanging from Feng's grip to his amazement he finds a Co-anda-19 tablet in his pocket. By luck it had fallen out of the tube he had lost. A sigh of relief crossed the face of Richards, which seemed to wind Feng up into more fury as he squeezed his grip tighter. He wanted Richards to talk more.

He realised if he took the Co-anda 19 drug, it would not necessarily release his imagination more so, it would just make his thought more fluid and lucid straight away under the stress he was suffering. He remembered Luigi saying to him as a young boy. "The power of the creative mind will come from within when you least expect it. It will most likely happen when you are focused and determined whilst at the same time demonstrating self-belief through your own spirituality."

Recognising that his mind driven by spiritualism is more powerful to achieve than relying on an enhancing drug, then suddenly it also dawned upon Richards that he still mentally had the upper hand even though physically he hadn't. He had a moment of clarity, which made him think Feng was correct. If he divulged more information about his hidden ideas to Feng, he would complete his duty in influencing someone to carry out his vision, a mark of true leadership. Whether he was around to see it being delivered or not, it didn't matter whether it was himself or someone else delivering it on his behalf. If he was to tell Feng more, it was Feng's choice to act upon delivering it or not. So, whether Feng knew more or not, it wouldn't make a difference to delivering it incorrectly. As it was a philosophy, not a design, that was at risk of being delivered incorrectly. Which wasn't a risk, as a philosophy would either be delivered or not. Therefore, Richards could take pride in influencing the android to carry out his work for him, if the philosophy had been delivered by Feng or by anyone else. It was about doing good for society, not for the glory of one person or organisation.

And then just like that a moment of clarity had struck down upon him, as Feng continued to press tightly around his throat as he felt his limp body lift off the floor like washing being hung out to dry. A sudden flashback to remembering the words of Merisi, *"The chosen one not only has to use the SATOR SQUARE to believe in the experience of time travel but understand what the SATOR SQUARE code personally means to them. So that they can create a path for us to follow, it is down to you to crack your own personal code, in relation to the SATOR SQUARE's code."*

CHAPTER 19

A FLASH OF BRILLIANCE

"**O**kay, Okay," Richards cried, as he grimaced with the pain from Feng pinching his neck. "I have seen enough of the future, to know a design philosophy I had already been working on devising, to be used in the present, is the correct way forward for the future prosperity of land resource management."

Feng released his pincers around his throat and lay him down gently. He knew Richards wanted to talk.

Richards immediately placed a Co-anda 19 tablet in his mouth as he lay in a heap. He realised it was too late to identify Feng's disguise before Feng revealed his true identity, and the drug wasn't needed as a vaccine against being drugged or given a virus by Feng, as Feng had already downloaded his ideas from his head. Instead, he decided to use the drug to expand his imagination in his hour of need, to try to tap into developing more new ideas inspired from his journey to the future that Feng would not know about.

"I have created a new type of farming, it's called – The five pillars of Penatgonal Farming. Firstly, eco affordable housing built from the onsite sustainable tree farming. More and more people can't afford to buy their own property and the paradigm shift is to live and work more rurally in the future as rural farmland offers firstly more affordable housing, powered by on site solar, enclosed by the tree planting and re-wilding that surrounds it."

"Secondly, tree planting and Re-wilding, with sustainable tree farming for sustainable timber-framed buildings to supply the onsite Eco- affordable housing, and for the trees to help reduce a local area's CO_2 emissions through increasing that given area's photosynthesis, by more trees being located in that given area. A programme of tree planting and new afforestation planted across farmers' land that has never had trees, with species native to the specific area, to provide flood protection and increase biodiversity, and restoring ecosystems."

"Thirdly, renewable power generation - large scale solar farming; cluster site connection with Electric Vehicle charging used to provide renewable power and transport to onsite Eco affordable housing."

"Fourthly, renewable heat generation from a Ground Source Heat Pump station to feed heating and cooling to on-site energy centre/ Biomass Combined Heat and Power system generated from recyclable wood chip sourced from the on-site sustainable tree farming. Heat energy transferred from the on-site Energy centre, collecting the heat generated from the GSHP and Biomass CHP system to on site Eco Affordable housing. A new on-site substation will be required to create a Microgrid named - MACRO-CHIP. Which stands for the 'Management And Centralised Renewable Organisation of Combined Heat & Integrated Power,' relating to availability of supply and demand on the local community energy network. MACRO-CHIP is the microchip of infrastructure and will be developed through my new brand of re-thinking, innovative renewable solutions for the future which I have named 'Infra-Green' inspired by the word Infra-Red, due to incisively looking at the detail of an idea whilst considering the big picture to execute the innovation. With this insight into the future of how technology has not fulfilled its full potential, I realise that there is plenty of existing technology already on the market in the present. By looking at delivering these technologies in a more innovative way, they will deliver betterment for humankind. It's the artful way of understanding how to get the best out of existing technology in a way that maximises its use through the science, its design, the economics, and the environment as one whole integrated and innovative solution."

"Lastly, vertical farming – two or three tier farming, through growing renewable transport fuel such as biodiesel from rapeseed oil, besides growing conventional crops. We are also living in a growing populated planet which needs sustainable food source. Providing an increasing diverse source of vertical farming, where farmland can be maximised for growing two or three times more food produce from the same land space due to having three or four tiers of food or/ and bio-fuel produce growing, or the same amount of produce growing within half or a third of the land space to provide more space for the other pillars of farming."

Feng, still attentive to Richard's explanation, he continued to tower over him, nodding his head whilst clenching his machined fist.

"My philosophy was complete after I was inspired from what the SATOR SQUARE means to me personally, on my own personal journey to completing my philosophies and designs. So I have named the concept- 'Arepo's farm.' The 5 pillars are signified as the five mystical palindrome words featured in the SATOR SQUARE and summed up symbolically as 'Arepo's farm,' to adapt to rethinking how a sower utilises their land resource in future generations. The sower named Arepo holds with effort the wheels. 'SATOR - AREPO - TENET - OPERA – ROTAS'... This can be translated as follows: The sower, with his plough, holds the wheels with care, or - according to other interpretations - 'The sower Arepo leads the plough with his hand.' Therefore, Sator means the sower, Arepo is the name of the sower, tenet is the plough, opera is the care and effort taken by the sower, and rotas is the wheel representing the new technology used to sow the efforts of the yield farmed in any new future generation, to utilise optimum use of land resource."

"Conceptually 'Arepo's farm,' is a philosophy which creates

the vision for the human race needing to Rethink farming in the future, to adapt to using new technology, and new caring ways of creating different yields to live sustainably and help climate change in the future."

"This five-pillar system will do for the future of agriculture, what the microchip has done for computers. The new strategic philosophy is a philosophy called '*Agri*-vision.' The main vision and goal in mind is to achieve the Management And Centralised Renewable Organisation of Combined Heat & Integrated Power. The management and centralised renewable organisation of combined heat and integrated power is to be consumed on site by the eco- affordable housing units, from the heat generated from the biomass sourced from the sustainable tree farming, and the power created from the solar farming. Providing sustainable farming for the onsite housing units from the produce and the bio-diesel fuel yielded from the onsite vertical farming."

"Enough, enough...Doctor! I have heard enough. Don't confuse my memory cells with too much information, we are not built like you humans!" as Feng sniggered to show a slight smile on his robotic face.

Richards looked relieved as he had last broken Feng's defence of evil tirade.

"So, what inspires you to come up with your visions, Doctor?"

"Less talk and more focus and concentration in thought," he whispered, trying to block out the pain, "I now realise what Plato meant, when he said – 'Wise men speak because they have something to say. Fools because they have to say something."

"Can't disagree there Doctor, but you better save your breath, it may not last."

"And deep thought, is much more powerful than spoken words. The pen is mightier than the sword," Richards added, "Let go of what's behind you and use your purpose to pull you forward. Use your weakness to mentally drive your purpose physically."

"To become elite in what you do, you have to develop a fantastic obsessive compulsive need to go beyond... if you don't have that, you won't be able to make the sacrifices necessary to achieve that level of success. Obsessive hard work equals success. Perfecting skill to be good at something and to be the best, equals success. The will to demonstrate competitiveness and determination and show these actions of determination speaks louder than words, equals success."

"What is your mind running away from Doctor?"

"To be able to achieve success through the abilities, one has in Determination, Drive, Work Ethic, and the natural desire to Want to be good at something."

"You have suffered a loss to be able to drive you to your success. Your motive has been both **spiritual in body and mind** to achieve your success, Doctor."

"Being good at ideas generation is down to my determination, desire to be an ideas person, and my work ethic."

"It seems that I have underestimated your tenacity doctor... although your time has run out." The Doctor's eyelids began to fall, as daylight began to fall and darkness drew close.

Feng started to lose interest, turned and walked away with a copy of the artefacts in his grasp, as Richards' unknowingly fell into a trance whilst whispering, "Everyone's personality is different, if we all had the same personality it would be a boring world."

"Make no mistake, in order to get to those extremes of success, you have to be overcompensating or super-compensating for an extreme 'need.' You 'Over do!' 'Determination' is the key to success through initially demonstrating mental 'Focus.' Make a goal then focus on achieving it."

"I admire your stamina and perseverance, to the end, Doctor."

"To achieve greatness in many areas, you have to be fantastically gifted and self-motivated and disciplined with immense organisation and leadership. Being extremely flexible, adaptable and versatile in your mindset, to be the best you can be, all the time!" Richards continued.

CHAPTER 20

FLASHBACKS AND PREMONITIONS

Feng sped off in the direction of the capsule and was gone. Richards let out a piercing scream, as he suddenly became aware of the pain in his lower limbs. Before he fought passing out, blood ran freely from his wounds, and soaked into the dew-moistened turf beneath him. The images of Shui Feng leaping into the white sports car outside the restaurant in Milan, and of what he thought was Grazia springing into the seat of the capsule in Beijing, flashed through his mind. Then his mind skipped to a flashback image of the metropolis drawing he had seen back in Grazia's flat in Milan. He began to think how the metropolis scene could be adapted to his own future metropolis scene. Showing imaginatively a city used not only using their sports stadiums but vacant buildings as an energy plants generating onsite renewable electricity from solar roofs and small wind turbine array and Ground Sourced Heat Pumps for centralised district power and heating distributed out to rural areas where people now work and live predominantly. A paradigm shift in replacing office space in the city once used for commerce was replaced with industrial renewable energy generation operations, brought about shortly after the pandemic in 2019.

It suddenly dawned upon him, as he lay dying, so far away from home, it's not just the stadiums that can be used for innovative renewable energy generation, but vacant commercial and residential buildings in cities can be transformed as well.

There is more demand for cleaner energy more than ever after 2019 with the demand for cleaner air to help combat against the growing number airborne viruses besides increased CO_2 emissions. Increases in CO_2 emissions have proved to increase the spread of more viruses more rapidly due to hindering air quality in the first instance which will cause airborne viruses to linger longer.

He began to feel his daydreaming become more lucid due to his current anguish impacting his mental health condition. Was he going to survive... was he going to make it?

His thoughts raced while he drifted in and out of consciousness.

Oh my God, he thought, it's all starting to make perfect sense. Renewable energy plants replacing the operational use of commercial and residential use in existing buildings within cities, as a quicker way to legally hit the target of CO_2 emissions by 2050, and a more economical way to save cost of energy and is a new way to make money by transforming the financial sector of cities into a new urban energy generation sector and store the energy and sell it to other nations worldwide,

It's proven to create higher amount of energy yield through better efficiency of energy generated from integrating wind energy on buildings by using the Coanda effect. It's too costly to generate renewable energy too far away from end user with green belt onshore and offshore wind farming,

The negative environmental impact is reduced significantly using cityscape renewable generation. The cost of travel and inconvenience of affecting people's balanced lifestyle having to endure long daily commuting will become a thing of the post 2019 pandemic. This all points to more expansion of living and working in more rural settings.

A sudden vision of being awarded the Nobel prize for climate change flashed past his mind.

Richards gradually drifts off into another flashback memory, remembering Father Luigi and what he had told him about the SATOR SQUARE, as it dawned on him, the SATOR SQUARE was not only named after the spiritual and scientific mode of time travel, but it also unlocked the secret design philosophy of saving the planet from Climate Change, and Nuclear Disaster to avoid Armageddon.

He flashes back to his conversation about SATOR SQUARES with Father Luigi in the coffee shop, *"So, what did they use these things for?" he asked.*

"Most often as a warning to people who had something to protect," replied Father Luigi.

"You mean their fortune, or their wealth…"

"Maybe? But not just material wealth. Very often, it was the protection of knowledge that was most important for people. It might be some valuable piece of military intelligence, or a commercial or industrial secret that enabled one city or community or religious order to maintain dominance over their rivals. I am sure that the ancient world was, and still is, littered with stashes of manuscripts and other sources of knowledge and wisdom. It was particularly commonplace in ancient China, for example. Prior to going to war with a rival warlord, or maybe when defeat appeared to be inevitable, the practice was to dump all evidence of the knowledge that had helped each society to prosper, in the deepest lake or at the bottom of the deepest mine."

"As I explained to you, in the cathedral, the meaning of the SATOR SQUARE is not known for certain, and scholars have suggested a number of translations from Latin, but the general message from the five words that make up the Square appears to be that 'the sower controls the turning wheel.' In one sense, it is a prophecy suggesting that the originator of a resource, or an initiative, or an idea is the one who is in control of its outcomes, good or bad – but, at the same time, it is a warning that there will be others who will try to wrest ownership from the sower. The rationale behind stowing away a cache of knowledge and other valuable information is

that it can be re-used when the cycle of power comes back in favour of a particular society or sect. Even if all the artisans and scholars had been murdered, their knowledge and know-how can be rekindled. The pursuit of these less lucrative treasures is a very under-rated branch of archaeology."

"Maybe the person who gave you the card bearing the SATOR SQUARE is trying to warn you to protect what is yours."

He then remembers feeling the warmth of Father Luigi's blood that gurgled quietly from his mouth, seeping into the fabric of his shirt and his trousers. Then his friend's body stiffened slightly. Then all was completely motionless. He remembers feeling a paranoid sense of being hunted and of being followed that had haunted him during that evening returned, as he allowed Father Luigi's lifeless form to fall onto the tiled floor of the café. They had got to his friend and lifelong mentor, and now they had got to him. It was now time to mourn the loss of Father Luigi, and the disappearance of Grazia. He felt the most loneliness he had felt so far on his journey, as he lay there dying from his injuries. He felt he wasn't going to make it and remembered Luigi's wise words spoken to him as a child, "Better to have tried, than to not have tried at all."

He then suddenly flashed back to the conversation he had with Merisi about his conversation he had with Luigi in the cathedral, "... but Father Luigi never mentioned anything about time travel," he said.

"I would not expect him to know about that, Dottore," said Merisi, "It is something that was known to a select few among the early Christians, in Europe – and by a similar number of holy men elsewhere in the world - and there are even fewer who are now acquainted with the true purpose of the SATOR SQUARES." He knew this was not true, as after speaking to Father Luigi in the coffee shop, he did know about its connection with time travel. Did Father Luigi know more than he was telling Richards? Did he know the true secret of how the time travel portal was discovered or created by man, which has been hidden by the church? He remembers saying to Father Luigi, "So the SATOR SQUARE is a code?"

"Yes! It concealed its meaning from nonbelievers, while

revealing its meaning to believers," responded Father Luigi, "It functioned as a kind of key for those in the know."

Then, as he gradually loses consciousness, in his mind's eye, he views himself on his back, moving quickly and looking upward.

His senses and vision are obscured, and he cannot see or hear anything for a lingering moment or two. A dull, intermittent sound draws close, gathering volume, louder and louder as he slowly came to his senses. The contrast between darkness and light fades in and out, as his vision blurs and flashes on and off. He can just about make out swishing blades that revolve and flicker above his head in his distorted field of tunnel vision. He can't quite tell if the muffled sound and vision are that of a helicopter rotor blade. His mind's vision fades out into total darkness and can hear himself say:

But what are the blades?

Where are they?

Why are they there?

Suddenly, I feel my entire torso jolt.

He hears the words... "Let her go, she is gone, she is gone."

ACKNOWLEDGMENTS

Thank you to Jim Arrowood who as my testimonial reviewer was instrumental in identifying and proposing editing requirements for book 2. Thank you to Christian Griffiths as a testimonial reviewer who also identified further editing required.

Thank you to Dr. Philip Barham who again supported me in writing some of book 2. Without his contribution as a ghostwriter, this story would not have been completed. Thanks again to Dennis Woods from Koehler Books for finding an interest in my story and recommending it for publishing. Most of all, thank you to my wife, Jacqueline, and my son, Ethan, for their continued full support, belief, and understanding to help me believe in myself and improve the initial final draft of this second book inspired by current observations and personal experience of contracting the Covid-19 virus. Whist the events and characters fortified in the story of book 2 are fiction, they have been influenced by my own semi-autobiographical research, observations, and experience over the years and currently. The first book was an allegory of capitalism, and corporate and international politics. Whilst this second book is an allegory of leadership, inspired from my own personal leadership experience throughout; my childhood and as an adult, and experience of leadership throughout business and construction. I am currently a lecturer and examiner on Leadership, on the International MBA course at Bedfordshire University, in England.

www.ingramcontent.com/pod-product-compliance
Lightning Source LLC
Chambersburg PA
CBHW040855210326
41597CB00029B/4852